D1015094

THE
PROJECT
MANAGER'S
DESK
REFERENCE

A
Comprehensive
Guide To
Project Planning,
Scheduling,
Evaluation,
Control & Systems

James P. Lewis

IRWIN
Professional Publishing®
Burr Ridge, Illinois
New York, New York

ISBN 1-55738-461-4

Printed in the United States of America

BB

4   5   6   7   8   9   0

This book is dedicated to Stuart F. Meyer,
W2GHK, my first mentor

Contents

v

Chapter 2

Chapter 3

Section Two
Project Planning   37

Chapter 4

Chapter 5

Chapter 6

Chapter 7

Section Three
Scheduling with CPM and PERT   91

Chapter 8

Chapter 9

Chapter 10

Chapter 11

Section Four
Project Control and Evaluation

Chapter 12

Chapter 13

Section Five
Key Factors for Success

Chapter 14

Chapter 15

Chapter 16

Chapter 17

Section Six
Progress Payments   243

Chapter 18

Section Seven
Managing Quality Improvement Projects   289

Chapter 19

Section Eight
Tools for Project Managers   305

Chapter 20

Chapter 21

Chapter 22

Chapter 23

Chapter 24

Section Nine
Problem-Solving in Projects   343

Chapter 25

Chapter 26

Chapter 27

Chapter 28

Section Ten
Improving Communications in Projects 381

Chapter 29

Section Eleven
Special Topics in Project Management 395

Chapter 30

Chapter 31

Chapter 32

List of Figures

Preface

Project management is finally coming into its own. Tom Peters, in his 1992 book, *Liberation Management*, says that in the organization of the future *most* of what gets done will be done in projects. I agree. But the future is now. The need to manage projects efficiently and effectively is upon us. Those organizations that take project management seriously as a discipline, as a way-of-life, are likely to make it into the 21st century. Those that do not are likely to find themselves in good company with dinosaurs.

When I first began working as a project manager around 1965, there weren't many sources of help on how to manage projects. I looked for a book that would offer me *practical* help. Unfortunately, I found many of them to be too theoretical, and that situation seems to be true even now. So it occurred to me that there must be project managers like myself who would like to have a book to which they can turn for real-world answers to their questions. A book that would be a ready-reference. In short, a handbook.

I must confess that I love handbooks. I love being able to look up almost any subject and find an answer to it in my many handbooks. That was the idea behind this one.

Thus, if you are trying to evaluate or locate scheduling software, language training, degree programs, references to books and articles on project management, you will find something on it in this book (the bibliography contains a listing of hundreds of books and articles written on the subject). There are checklists, forms, listings of consultants, associations, and other material of interest to project managers. I did leave out the kitchen sink, but just about everything else is in there.

While I do not delve into some of the esoteric areas in which projects are run, I think the book will be useful to almost any project manager, regardless of discipline. It is my hope that you will find it so useful that it remains on you desk for ready reference—thus the title *The Project Manager's Desk Reference.*

It is my conviction that the tools of project management are much like mathematics—they can be applied to everything. Two times two is four, whether you talk about apples or supernovas. The approach to running any project is the same. For example, if you consider surgery, you can think of an operation as a project. Every one is a little different. And every one must be approached in a similar way. The physician must *diagnose* the problem, *plan* the steps to take, *schedule* the operation, *implement the plan,* and *monitor and follow-up* to ensure that the operation will be a success. So, too, with any project.

While this is not an academic book, I believe it will be useful for training programs on project management, and I suspect that we are going to see a tremendous increase in such training in the next five years. You will note too, that schools are beginning to offer degrees in project management.

A book of this type certainly is not possible without the combined efforts of a number of people. My wife, Lea Ann, has spent countless hours preparing nearly 140 first-class figures to illustrate the text. She has given me invaluable help in making the book a quality product.

I appreciate the contributions of Quentin W. and Quentin J. Fleming. Chapter 18 is taken from their Probus book, *A Probus Guide to Subcontract Project Management and Control: Progress Payments.* This chapter not only covers progress payments, but also explains the C/SCSC or earned-value concept very well.

Dr. David Antonio, of the University of Wisconsin-Madison contributed a chapter which details the results of a research study on the effectiveness of measuring project success using the Project Implementation Profile developed by Dennis Slevin and Jeffrey Pinto. Chapter 14 is also written by them. It was formerly published in the *Sloan Management Review*, and is reprinted here by permission. I appreciate the contributions of these authors.

I am grateful to Applied Business Technologies for making available to me a copy of Project Workbench™ for my use in the scheduling section of the book. They have also been very supportive in my efforts to document my books with their software.

I would also like to thank all of the folks at Probus for their hard work in producing this book. To Pamela van Giessen, my acquisitions editor, I say thanks for your patience. To Kevin Thornton and the production staff, I also thank you for all the work that went into making this a quality book.

As for the deficiencies, they are, of course my own.

A final word. I intend to update this book periodically, with the goal of making it *the* project manager's reference book. I would appreciate feedback and contributions from readers. The contributions need not be formal chapters. I would welcome hints, techniques, and other suggestions which you think would be helpful to other project managers. Please send them to me care of the address which follows. You will be credited with each and every contribution.

James P. Lewis
302 Chestnut Mountain Drive
Vinton, Virginia 24179 U.S.A.

Section One

Introduction to Project Management

Chapter 1

Introduction to Project Management

The Changing Landscape

The quality revolution that began in the 1980s in the United States is slowly bringing about a management revolution as well. The old paradigms, which have influenced the thinking of most U.S. and many European managers, have proven increasingly ineffective in meeting both local and global competition. Still, as is always true of paradigms, the old ways linger on. As Kuhn (1970) demonstrated in his now famous work, *The Structure of Scientific Revolutions*, all models give way to new ones slowly and sometimes painfully.

> par • a • digm: a model of reality

Paradigms

Nevertheless, the changes are taking place. As the old paradigms give way to the new, we now have a new language of management, which is derived from those new practices. And yet, the practices are not entirely new. They were originally suggested by our own gurus in the 1950s and 1960s and taken seriously by the Japanese and practiced. Two of those gurus were Deming and Drucker.

When you read what those men were saying over thirty years ago, it sounds very *modern,* and it should, since their ideas are now central to the philosophy of the Japanese and practitioners of quality in the U.S. As Drucker (1956) wrote in *The Practice of Management,* "The primary purpose of a business is to create a customer."

Yes, we all know that the primary *motive* of a business is to make a profit, but as someone has said, the customer is the only person who will give you the money that becomes your profits. Your stockholders don't do it. Banks don't do it. Only the customer creates profits. So the first order of business is to find someone who is willing to be our customer. Without customers, we are dead. Period.

The Japanese read that and took it seriously. I recently heard someone say that the Japanese read our books on management and were naive enough to believe that we really did what it said in them, so they did what the books said and sure enough, it worked! What we are now doing in the 1990s is rediscovering what our experts were saying back then.

Fine. So what does all of this have to do with project management? The answer is—plenty.

Alvin Toffler (1971) wrote one of the first books to make us realize the accelerating pace of the world in which we live. In *Future Shock* he pointed out that change is increasing at a dizzying rate, and suggested that we had to learn to adapt.

Future Shock is now *present shock!* A few examples will illustrate.

Until recently, the American auto industry took six to eight years to develop a new model car. The Japanese do the same thing in three years. And even the upstart Korean company Hyundai, which did not make cars when the GM Saturn project was started, got several new models into the American market before the Saturn hit the streets, four years behind schedule (Dimancescu, 1992, p. 55).

In a recent film entitled *Speed Is Life*, Tom Peters talks about the need for quicker product development. It once took four years for a Pennsylvania division of Ingersoll-Rand to design a new hand-held grinder, a small device for removing burrs from metal parts. By applying cross-function management, they recently got it down to one year.

The new Boeing 777 aircraft, which will contain several *million* parts, is scheduled to be in production in five years, from start to finish, which is one and one-half years earlier than happened for the 767 (Rosenblatt & Watson, 1991).

It can no longer be business-as-usual.

Project management methodology has been fairly static since CPM (Critical Path Method) and PERT (Performance Evaluation and Review Technique) scheduling were introduced around 1958 and the Earned-Value measurement concept was developed in 1963. The problem is, the old tools are proving to have shortcomings. To be more blunt about it, they don't work in a lot of situations.

There are at least two primary reasons for the shortcomings. One is that scheduling methodology does not handle *concurrency* very well. The second is that earned-value analysis does not provide a proper "picture" of the health of a project, because it focuses only on financial measures and leaves hanging a core principle of what business must accomplish—the meeting of the customer's needs!

This does not mean that we should "throw out the baby with the bath water." The old tools have their place. But we must augment them with modern methods to ensure that projects are managed with the same customer focus as a business.

In his book, *The Seamless Enterprise* (1992), Dan Dimancescu has a term for *customer* that really captures the essence of the word in its broadest meaning. He calls the customer the *next in line*. Whoever is the immediate recipient of whatever you do is your customer. And the focus of all business must be to see that the customer's needs are being met.

This is contrary to what many of us have done for a long time. In product development, the discipline in which I spent 15 years, it has been common practice to design a product without talking to the ultimate end customer about his or her needs, then "throw it over the wall" to manufacturing and expect them to build it, even though it was designed without considering their capability to make it.

The net result has been costly. The Edsel and the PCjr are examples of products that were someone's bright idea, but that the customer did not accept. And as much as 30 percent of product development cost—the total cost to get a product into manufacturing—is often rework. Dimancescu quotes the chief engineer of Boeing's 777 aircraft as saying that this amounts to one of every three engineers spending full time redoing what the other two did wrong.

We can no longer afford such waste.

Still, as I have said above, old habits die hard, and there is a tendency to hang on to the old ways of managing projects because they are comfortable, and worse yet—they may have worked in the past.

In fact, I believe that there is nothing more dangerous to the survival of any business than success, as it often leads to complacency. The response given to suggestions that we should change is, "Why should we change? We're doing fine."

To help see how this works, I use a model of change that I first learned from Marvin Weisbord's book *Productive Workplaces* (1987). He said that people live in various rooms of a four-room house. The house is shown in Figure 1.1.

When things are going well, they live in the room called *contentment*. At the first hint of trouble, they move to the room called *denial*. "We don't have a problem," they say. At the individual level, this is the response of the person with a drinking problem. "I don't have a problem," says the person. "I just drink socially. I can quit anytime I want."

At the corporate level, we say it is "unfair Japanese competition," or we blame the government for our loss of markets. It can't possibly be us.

However, the evidence continues to build and we finally have to accept that *we do have a problem after all!* That causes a move to the *panic* room.

Figure 1.1. The Four-Room House.

When a person or organization moves into the panic room, they frantically search for a quick fix for their problem. This is what causes American companies to look for the "program-of-the-month," the latest gimmick, which they hope will solve their problems. They try something, it doesn't offer immediate relief, so they abandon it and move on to something else.

What is important to note is that, if you can't get a person or an organization out of denial, there is no hope. Once you get them out of denial, you then have a chance to move them into renewal, but only if you can keep them focused on one program long enough for it to work.

Finally, they can move back into contentment.

However, as Dr. Deming (1982) has argued, no organization can be satisfied to stay in contentment too long. They must constantly be searching for ways to improve. Constant improvement is the byword of business—actually of *any* organization today. If the organization does not improve, it will eventually die or be displaced by another company that does things better.

In terms of the four-room house, what this means is a constant moving back and forth between contentment and renewal, without the pain of having to go through denial and panic anymore. To me, this is the essence of where we want to be. We want the organization to be what has been called the *learning organization* by writers such as Peter Senge (1990).

Again, this applies to *project organizations* as well as to businesses as a whole. A project team is a microcosm within the larger whole. The same practices that apply to the "parent" must be applied to all of its parts.

In systems terms, the word *organization* has the same root as the word *organism*. We know that if one part of an organism is unhealthy, then the entire animal may suffer. Problems with one's teeth can poison the entire body. So, too, with parts of an organization. If project teams do not function properly, then the entire organization may suffer.

There is one significant point that should be made about this. It is that we must be careful about *suboptimization*. To use the analogy of a sports team, it does no good for one player to be a superstar if the total team cannot play together. This has been one of the problems with companies in the past. Only if all parts work in concert can the entire organism function in its overall environment. This lesson must be applied to project teams.

The New Face of Project Management

So what does the project team of the 1990s look like? How does it function? How is it different from those of the past?

The key component that governs the new face of project management is the concept of *concurrency*. It has been applied under the banner of concurrent engineering. The problem is, many projects are not engineering oriented, which leads to the belief that concurrent engineering has nothing to do with those projects. However, it is the concept of concurrency that is important, and it is my conviction that this concept must be applied to *all* projects, no matter what kind.

One of the best definitions of concurrent engineering is offered by the Institute for Defense Analysis in their report R-338 (Winner, et al., 1988, *The Role of Concurrent Engineering in Weapons Acquisition;* see references). This definition provides the basis for understanding concurrency:

> *Concurrent engineering is a systematic approach to the integrated, concurrent design of products and their related processes, including manufacture and support. This approach is intended to cause the developers, from the outset, to consider all elements of the product life cycle from conception through disposal, including quality, cost, schedule, and user requirements.*
>
> *Concurrent engineering is characterized by focus on the customer's requirements and priorities, a conviction that quality is the result of improving a process, and a philosophy that improvement of the processes of design, production, and support are never-ending responsibilities of the entire enterprise.*

Following this definition, I want to offer a new term, *concurrent project management,* to convey the core idea. I will use the word *product* to mean whatever a project team produces, whether it be sofware, hardware, information, services, etc. Following is the definition of concurrent project management:

Concurrent project management is a systematic approach to the integrated, concurrent design, development, or construction of products and services, including their related processes such as manufacture, delivery, and support. The project team will consider from the outset all elements of the *product* life cycle from conception through disposal, including quality, cost, schedule, and user requirements. It is characterized by a focus on customer requirements and priorities, a conviction that quality is the result of improving a process, and a philosophy that improvement of all processes is a never-ending responsibility of the entire project team.

What this definition says is that there will be no more development of product without consulting the end customer. The "next-in-line" will also be part of the development effort so that his or her needs are addressed from the outset. The project team will cut across the functions of an organization to coordinate the total effort so that total organization performance is optimized, not just individual units within the parent.

Dimancescu calls this *cross-function management*, and points out that it means not just a multidisciplinary team getting together and doing a job, but managing the entire project so that it cuts across the functions (chimneys, to use Dimancescu's term) of the organization to achieve overall optimization.

This, I believe, is the future of project management. In this introduction, I cannot cover the subject in depth. That is done in various chapters throughout this handbook. I hope, however, that this brief introduction will stimulate readers to rethink the face of project management and develop their own unique practices.

Chapter 2

Overview of Project Management

Concepts of Project Management

This chapter will establish definitions of some terms and introduce general concepts about project management.

What Is a Project?

In recent years students of organizations have found that as much as 50 percent of the work done in typical organizations is done in a project format. This makes project management an important discipline and one that is receiving increasing attention. For that reason, it is important to have a clear understanding of how project man-

---

agement differs from more general forms of management. This obviously includes a better understanding of the difference between projects and other kinds of jobs. Following are some examples and counter-examples of projects.

> **A project is a one-time job that has defined starting and ending dates, a clearly specified objective, or scope of work to be performed, a pre-defined budget, and usually a temporary organization that is dismantled once the project is complete.**

EXAMPLES

☞ Developing a new product or service

☞ Building a bridge, house, road, runway, or other structure

☞ Writing software

☞ Installing a new manufacturing process, cell, or assembly line

☞ Publishing a book

☞ Developing a new marketing plan

COUNTER-EXAMPLES

☞ Processing insurance claims, orders, or invoices

☞ Manufacturing widgets

☞ Cooking in a restaurant

☞ Driving a delivery truck over the same route every day

☞ In short, anything of a purely repetitive nature

Another definition of projects that has considerable merit is the one offered by Dr. J. M. Juran (1989), the quality expert. His definition is presented in the box as shown.

> **A project is a problem scheduled for solution.**
> **—J. M. Juran**

As this definition indicates, project management is problem-solving on a large scale. One of the common causes of difficulty in running projects is that insufficient time is spent at the beginning of the job defining exactly what problem is to be solved by the project. This can lead to the unfortunate situation in which the right solution has been developed, but for the wrong problem. Guidelines will be presented in the planning chapters of this book on how to avoid this problem.

Definition of Project Management

> **Project management is the planning, scheduling, and controlling of project activities to achieve project objectives.**

Project management involves three major activities aimed at achieving project objectives. These are called planning, scheduling, and control. Each of these activities will be treated in major sections of this book. For now, we will examine only the primary objectives that exist in all projects. These are listed as follows. We say that project work must be complete:

P ☞ At the desired performance level

C ☞ Within cost or budget constraints

T ☞ On time

S ☞ While holding the *scope* of the project constant, and

☞ While using resources efficiently and effectively

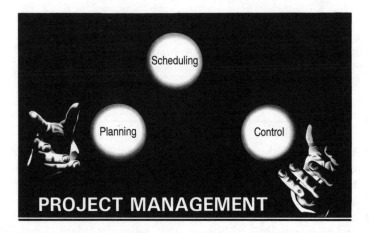

The first three of these are referred to as the **P**, **C**, and **T** aspects of project management. Some people call them *good, fast, and cheap*. (P = good; C = cheap; T = fast) The term *scope* refers to the magnitude of the job as well as certain other boundaries or constraints. For example, if a house is priced with the understanding that conventional plumbing fixtures are to be installed and the buyer asks for gold-plated fixtures, we say that is a change in the scope of the job and will result in a price increase.

A very important point:

> You cannot tie down all four of them simultaneously. If three are specified, the fourth must be allowed to vary.

Mathematically, this can be illustrated with a general equation as follows:

$$C = f(P, T, S)$$

In words, the equation says, Cost is a function of Performance, Time, and Scope. Generally speaking, the cost of the project will increase as P, T, and S increase, except in the case of trying to crash the project. Then, below a certain time allowed, costs will begin to increase because labor costs begin to escalate. This is shown in Figure 2.1. Note that costs reach a minimum and then rise again be-

Figure 2.1. Project Costs as a Function of Time.

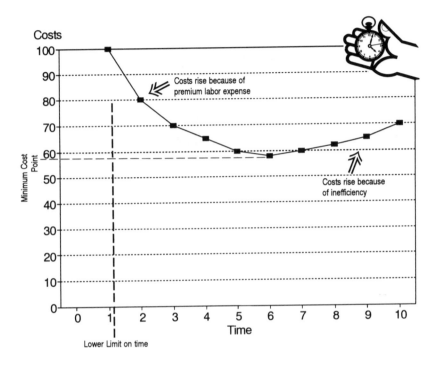

cause of inefficiency in how the work is done when it is strung out too long.

Another way to illustrate this is with a diagram, as shown in Figure 2.2. Note that Performance, Cost, and Time are related as a triangle, and that Scope increases simply cause the size of the triangle to increase. For a given scope, however, there will be limits to how long the sides of the triangle can be. For example, in the first illustration in Figure 2.3, the maximum length for the side representing cost is just under eight units. Once it reaches eight units, you no longer have a triangle, but a straight line.

Figure 2.2. Relationship of Performance, Cost, Time and Scope.

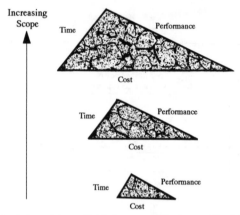

As project scope increases, Cost, Performance, and Time increase, although the
ratios do not necessarily remain the same as is suggested by this diagram.

There is also a minimum length that a side can have, as shown
in the second half of Figure 2.3. If the length of the cost side approaches two units, then the two sides overlay the time side, which
has a length of five units.

In spite of the fact that the four variables are interdependent,
managers often try to dictate all of them at once, then wonder why
they cannot be met. Note that one of the most common problems is
for the scope of the project to increase as time passes. People think
of things that did not initially occur to them. Or they did not take
enough time to properly define the problem being solved at the beginning of the job.

Unfortunately, the scope tends to increase in small increments,
rather than in large ones, making such changes a bit innocuous.
Such incremental changes are called *scope creep*. The difficulty is,
many people suffer from amnesia at the end of a project, which
means that:

> **You must control *scope creep!***

Figure 2.3. Maximum and Minimum Lengths for Side.

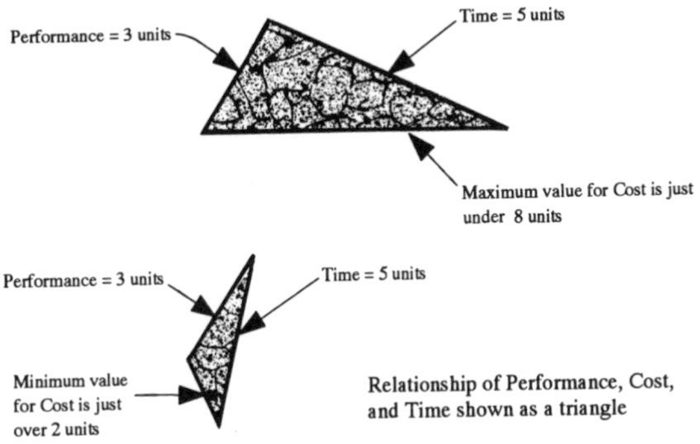

Performance = 3 units

Time = 5 units

Maximum value for Cost is just under 8 units

Performance = 3 units

Time = 5 units

Minimum value for Cost is just over 2 units

Relationship of Performance, Cost, and Time shown as a triangle

The Performance Objective

The often-forgotten objective in project management is the performance target. This target is not just a technical specification. It is a
translation of the customer's needs into performance criteria, and
that translation may be a technical specification. However, as was
pointed out in Chapter 1, concurrent project management requires
that the customer be a part of the entire process from concept
through completion, with *next-in-line* being the operational definition of customer. Failure to meet the needs of the next-in-line is a
violation of the practice of quality at the project level. The contemporary approach employed is Quality Function Deployment (QFD).

The Project Life Cycle

As a general rule, projects have a life cycle consisting of four to six
phases. For a six-phase model, the phases are called concept, definition, design, development or construction, application, and post-
completion. The character of the program changes in each life-cycle
phase (see the life-cycle illustration in Figure 2.4).

Figure 2.4. The Project Life Cycle and Effort Expended.

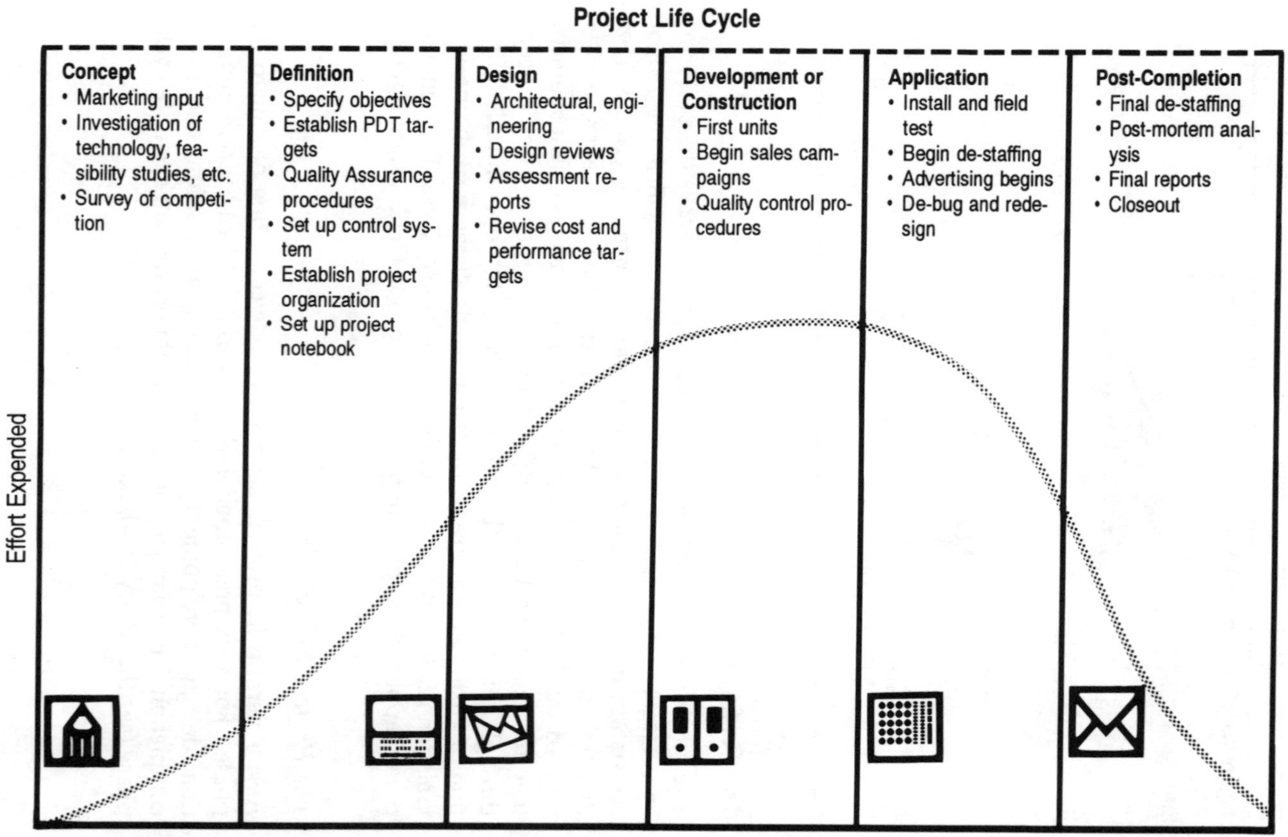

Project Life Cycle

Concept	Definition	Design	Development or Construction	Application	Post-Completion
• Marketing input • Investigation of technology, feasibility studies, etc. • Survey of competition	• Specify objectives • Establish PDT targets • Quality Assurance procedures • Set up control system • Establish project organization • Set up project notebook	• Architectural, engineering • Design reviews • Assessment reports • Revise cost and performance targets	• First units • Begin sales campaigns • Quality control procedures	• Install and field test • Begin de-staffing • Advertising begins • De-bug and redesign	• Final de-staffing • Post-mortem analysis • Final reports • Closeout

Effort Expended

There are two major pitfalls in the life cycle of a project. The first is that the concept for the project is accepted as the definition, leading to the problem mentioned previously, i.e., the right solution is developed for the wrong problem.

The second pitfall is at post-completion. Note that a postmortem should be conducted for the project. The aim is to learn what was done well in the job and what might need to be improved. However, this stage is often aborted. By the time the project is completed, people are ready to get on with something else.

More significantly, they may be reluctant to face the fact that some areas need improvement. Failure to face such issues may be caused by not wanting to embarrass anyone. Nevertheless, no matter how well a job has been done, there is always room for improvement, and a post-mortem should be conducted in that spirit. Naturally, any climate of blame or punishment simply increases the likelihood that no one will conduct an "honest" evaluation of a project.

The Project Management System

The project management system consists of seven components, as shown in Figure 2.5. If any one of these is not properly designed, then the management of projects will suffer. These system components are covered in various chapters of the book.

Figure 2.5. The Project Management System.

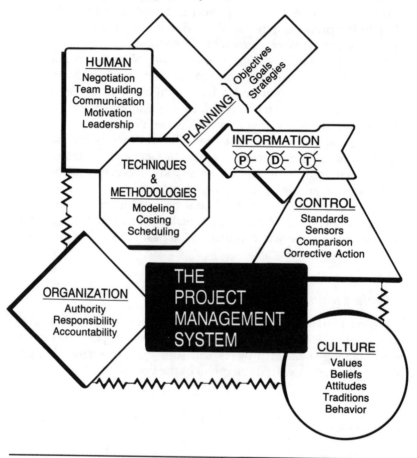

Chapter 3

A Model for Managing Projects

Overview of the Model

In Figure 3.1 is a flow chart that illustrates the steps that should be followed in managing a project. The steps in the model will be explained in detail in the appropriate chapters that follow. At this point, only a broad summary will be presented.

Steps 1 through 8 constitute the planning process, including project scheduling. Steps 9 through 16 specify the steps involved in monitoring and controlling progress. The model is designed to prevent the more common problems that seem to occur in projects, but it cannot capture the complexity of the entire process without becoming unwieldy.

Figure 3.1. General Model for Project Management.

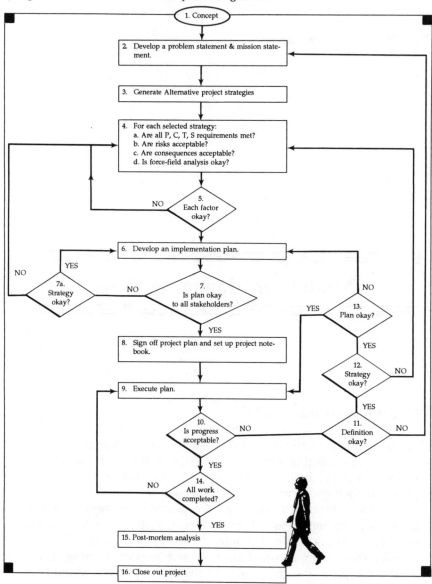

Step 1: The Concept Stage

As is shown in the project life-cycle model presented in Chapter 2, a project begins with a concept. Someone identifies a need for something. The problem is that a concept can be very vague. The identified need has not been thought through very thoroughly. For that reason, the concept stage is followed by the definition stage.

Step 2: Develop a Problem Statement

The next step is to develop a good definition of the problem that is intended to be solved by execution of the project. This is probably the single greatest hurdle that must be overcome in the entire process. Insistence on writing a problem statement is usually met with skepticism, the reaction being, "We all know what the problem is. Let's get on with it. This is a waste of time!"

Many examples have been described in the literature to show that this is often not the case. I will cite only one. In their book, *Breakthrough Thinking*, Nadler and Hibino (1990) tell about a company that received complaints from its distributors that it was sending them damaged goods. They hired

> Remember: A project is a problem scheduled for solution.

an efficiency expert to investigate, and she accepted the definition of the problem as offered—to reduce damage to goods. To solve the problem, the expert designed for them a computer-controlled conveyor to load trucks. She estimated that the conveyor system would cost around $60,000 per warehouse location, with savings yielding a payback in about eight months. Since the company owned 24 warehouses, the total investment was to be $1.44 million.

The vice president was inclined to accept the consultant's recommendation, but decided (perhaps for political reasons) to ask the internal industrial engineering group for a second opinion. The assignment was given to a staff engineer who was a recent college graduate. He studied the situation. However, rather than accept the definition of the problem as given, he asked a new question—what are we really trying to achieve? His answer was, "We are trying to find the best way of distributing our products to the marketplace."

Based on that problem statement, he completed the study and prepared a presentation for management.

When time came for the presentation, the vice president asked, "Well, do we go ahead and spend the $1.44 million?"

The young engineer responded, "No, sir. I think you should sell all the warehouses."

It turned out that he did not mean they should sell *all* of the warehouses—just most of them. The company ultimately followed his recommendation and sold all but a few regional warehouses, each stocked by air shipments directly from the company's manufacturing plants. Eliminating local warehouses simplified freight transfers so there were fewer physical handling points for each shipment, and consequently less likelihood of damaging goods. The solution saved the company hundreds of millions of dollars each year, and eventually forced their competitors to restructure along the same lines.

This example drives home a most important point about problem-solving: *The way a problem is defined determines the solution possibilities.* It is this fact that makes defining the problem correctly so important.

> **The way a problem is defined determines the solution possibilities.**

Chapters 25 and 26 present more-complete approaches to problem-solving methods that should be applied at this step.

Step 3: Generate Alternative Project Strategies

As the old saying goes, "There is more than one way to skin a cat." With most projects, there will be more than one approach that can be applied to achieve the desired result. For example, building a house can be done by starting from the ground up and constructing every single element at the site, or the house can be assembled from prefabricated parts. Further, it can be built entirely by one contractor or various parts can be subcontracted (for example, plumbing, wiring, roofing).

In technological projects, the approach may involve using proven technology for reduction of risk. Or "cutting-edge" technology may be used to achieve a competitive advantage, in spite of the fact that risk is increased.

The common approach to be used here would be to brainstorm a list of available strategies and then select one. Creativity-enhancing methods can be employed to increase the likelihood of developing a good strategy. Edward de Bono is considered by many to be the world's leading expert on creativity, and his recent book, *Serious Creativity* (1992), presents his approach in detail. The interested reader should consult that book.

Step 4: Select and Evaluate the Strategy

After a list of strategies has been developed, one must be selected. A strategy will be considered suitable only if it passes four tests. Answering these questions cannot always be done in a quantitative way, but the analysis is important in identifying potential problem areas before time is wasted developing a detailed plan.

Step 4a: All P, C, T, S Requirements Met?

The first question is whether the approach meets all Performance, Cost, Time, and Scope requirements. In other words, will it do the job?

Naturally, this is a judgement call, but experience can usually be used as a guide to answer the question. However, it may be necessary to do some broad-brush implementation planning before the question can be answered definitively. For that reason, when step 7 is reached, if the implementation plan is not okay, strategy must be examined to determine if it is acceptable, and if not, a new one must be developed and the process repeated.

Step 4b: Are Identified Risks Acceptable?

This step is intended to identify any risks that might cause the approach to fail. There are some managers who think doing a risk analysis is a bad idea because it causes people to begin *thinking negatively*, and they fear that the result will be to create morale problems. That might be the case if such an analysis is not done correctly. To simply ask, "What could go wrong?" and leave it at that would very likely cause people to consider the approach non-viable. However, I do not let people do a risk analysis in that manner. For every identified risk, I ask, "What might we do if it happens?" In other words, we must identify *contingencies* for every risk, if at all possible. This is illustrated in Figure 3.2 for a project to photograph archeological ruins.

I simply take a sheet of paper (or a flip-chart page) and divide it in half. Then I list risks on the left side and contingencies on the right. A few of these are listed in Figure 3.2, but the list is not meant to be exhaustive, only representative.

Occasionally, there will be no contingency for a risk. For example, when overhauling a large power generator, it may be necessary to remove the rotor, refurbish it, and then re-install it. These rotors are so large that they must be handled with a crane. One of the risks is that the rotor will be dropped, and there is no contingency, because they are specially built and there is not likely to be a spare sitting in a warehouse.

What, then, do you do? You manage that activity carefully to minimize the probability that you drop the rotor. The value in having identified this as a serious risk lies exactly in the fact that you know to pay special attention to this activity.

Step 4c: Are Consequences Acceptable?

In taking any action to solve a problem, peripheral effects may occur. These can be called *unintended consequences* of the steps taken. These unintended consequences result in new problems being created. The question is, then, can we live with those consequences. If not, then we must consider a different approach to the project.

As an example of unintended consequences, legislation was enacted some years ago in the U.S. to make streets more accessible to handicapped people. The law required that ramps be placed at street corners to make it possible to get a wheelchair across the street without having to negotiate difficult curbs.

This solved the problem for wheelchair-bound individuals, but created problems for blind people, who use their canes to search for the intersection by feeling for the curb—which has now been re-

Figure 3.2. Risk Analysis for Photography Project.

moved! It also makes sidewalks more hazardous for sighted individuals, because sudden drops in pavement level are not always
easily seen.

Another example of unintended consequences is the result of having a sale to reduce inventory levels. When the sale ends, some customers perceive the product to be of lower value than it was previous to the sale.

In a project context, we know that projects may have environmental impacts that are undesirable. Or in a product-development situation, pursuing "safe" technology may achieve a speedy time-to-market, but result in a competitive disadvantage. In addition, it may cause the organization to fall behind in the development of its technological capability.

As was mentioned previously, if such consequences are not acceptable, then a different approach (strategy) for the project should be considered.

Step 4d: Does it Pass a Force-field Analysis?

Of the four tests that a strategy must pass, the force-field analysis is probably the hardest to quantify. Nevertheless, it is well worth doing.

As shown in Figure 3.3, a force-field analysis looks a little like a risk analysis, but is very different. On the right side of the page are listed those forces in the environment that can be expected to *assist* in the implementation of the project, while on the left side are those forces that might *hinder* or *resist* its implementation.

Notice that these are mostly social forces. They result from the attitudes that people have toward certain approaches. For example, we sometimes hear people say in organizations, "We don't do things that way around here." If a project manager is attempting to run a project using an approach that a powerful member of management considers out of line with "how things are done around here," then it is likely that the manager's resistance may cause the approach to fail.

The factor labeled NIH means "not invented here." Sometimes individuals resist a particular approach simply because they did not think of it. Although this may be petty, such resistance can sabotage a project, and should not be underestimated.

Once both positive and negative forces have been identified, the method calls for the strengths of all forces to be estimated and tallied up, with the understanding that the sum of the positive

Figure 3.3. Force-field Analysis.

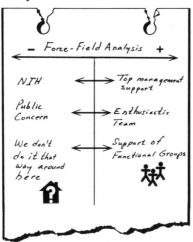

forces must be greater than the sum of the negatives or the approach will not work. The usual approach to measuring the forces is to rate them on a ten-point scale, then multiply by a weighting factor (assuming that they are not all of equal importance), and sum them.

In my opinion, this is an exercise in futility. It appears to give a measurement to something which I believe usually is unmeasurable. For that reason, I do not advocate the practice. Rather, I suggest simply trying to deal with the negative forces.

There are three approaches that can be used to deal with forces identified in the analysis. They are:

1. Strengthen the positive forces so they are definitely stronger than the negatives.

2. Find ways to get around the negatives.

3. Find some way to weaken or eliminate the negatives.

What we find people doing in case after case is choosing option one. They try to overcome the negative forces with stronger positive forces. This is in spite of countless experiences which teach that *the harder you try to overcome a negative force, the stronger it be-*

comes! This has been understood by systems theorists for years, yet it does not seem to be well understood by most of us. The "push-resist" interaction typifies all conflict and competitive situations, and the stronger one side pushes, the more the other side resists. All you have is escalation.

The most helpful way to deal with forces is to try to neutralize them. Find some way to make them go away. For example, if someone thinks the selected strategy is bad, ask that person, "What would I have to do to convince you that this is a good approach?"

There are two possible responses. The person can tell you to "forget it." You will never be able to convince him or her that the approach is sound. If so, you may have to forget trying to convince the person and decide whether to proceed or choose a different option.

However, I will always ask the person, "Are you sure there is *nothing* I can do? That's pretty heavy." Usually this gets the second response, which is for the person to say, "Well, I suppose if you can do this (they explain what it will take) I would be convinced." The nice thing about this approach is that you now know what it takes to "make the sale." You no longer have to hunt for the selling proposition. If you can do what the person suggests, you are "home free."

Step 5: Above Factors All Okay?

If you have passed each of the four tests, you can consider the selected strategy to be tentatively acceptable. It is still possible that you will find during implementation planning that there are problems with the approach and have to reject it, but this is not very likely. There is one caution: Very analytical individuals sometimes go into "analysis paralysis" at this step in planning a project. The purpose of these steps is not to identify every single risk or consequence that may exist, but to assess some of the more likely ones. Project managers may have to assert this a number of times with skeptical or negatively oriented members of the proposed project team.

Step 6: Develop an Implementation Plan

Up to this point, the planning process has answered the broad question of what strategy will be employed to manage the project. Now the strategy must be translated into specific steps to be taken to get the job done. These steps will define *what* is done, by *whom*, for *how long*, at *what cost*, etc. An overriding concern will be deciding how to translate customer needs into solutions. During this stage, a Work Breakdown Structure will be developed, a schedule using CPM or PERT will be formulated, resources will be allocated, responsibilities assigned, control systems developed, and so on. How this is done is the subject of Chapters 7 through 11.

Step 7: Is the Plan Okay to All Stakeholders?

A stakeholder is defined as anyone who has a vested interest in the project. This will include suppliers, contributors, customers, senior management, financial contributors, sometimes the community, and so on. In the case of contributors, we need to ensure that they can make their contributions when required at the desired level of quality. If the customer is considered, we want to be sure the work done will meet his or her needs. If the answer is no, then we must examine strategy (step 7a). If strategy is considered unacceptable at this point, the model routes us back to step 4, where a new strategy is selected and tested, and replanning is done.

If only the implementation plan is at fault, the loop routes back to step 6, meaning that the working plan must be fixed to the satisfaction of all stakeholders.

Step 8: Sign Off Project Plan and Set Up Project Notebook

Stakeholders indicate their approval of the plan by signature. This indicates also their okay for the execution phase to begin. Use of a notebook to contain all project documentation is recommended.

Step 9: Execute the Plan

At this point work begins. The detailed implementation plan is the guide to the steps taken during the execution phase. One pitfall that is sometimes observed in projects is that the plan is not followed during execution of the work. This is especially true when problems are encountered. It is tempting to forget the plan and just start trying to correct the problem. Note, however, that steps 10 through 13 are designed to handle problems.

Step 10: Is Progress Okay?

As work is performed, it is monitored. One of the principal tools for doing this is earned-value analysis, which is discussed in Chapter 13. However, it is important to remember that earned value analysis can be properly used only if the *performance* objective is being met. That is, work can be said to be on target only if it meets customer requirements. The fact that what has been done functions correctly according to a technical specification does not mean that the project is on target. The Edsel may have functioned correctly according to its engineering specs, but it was not accepted by the market. If the answer at this step is no, the model routes into control steps.

Step 11: Definition Okay?

This step ensures that we are still trying to solve the correct problem, rather than the wrong one. If the answer is no, the model routes all the way back to step 2, meaning the project must be totally replanned. This won't happen very often, but it must be considered as a possibility.

Step 12: Strategy Okay?

As in step 7a, it is important to ask whether an implementation difficulty is caused by a defective strategy. If it is, the model routes back to step 4, so another strategy can be selected.

Step 13: Plan Okay?

If the answer to this question is no, then we have to change the implementation plan. However, if the answer is yes, then the meaning is that the plan is not being followed. In many cases, the reason will be that insufficient resources have been provided. If not, then they must be forthcoming or the project will have to be replanned. Note that resources can be increased through the use of extra people or overtime.

Step 14: All Work Complete?

This just keeps looping back to the execution step, meaning that during execution we monitor progress, take corrective action when necessary, and so on. Once the answer is yes, we are ready to do step 15.

Step 15: Post-Mortem Analysis

Before the project can be considered actually complete, a post-mortem should be conducted. This audit is for the purpose of learning what was done well and what could be improved, so that progress can be made in future projects. See Chapter 12.

Step 16: Close Out Project

Final reports are written, the project notebook containing all documentation is placed in a central file, and the project is considered complete.

Section Two

Project Planning

Chapter 4

General Aspects of Project Planning

An Introduction to Project Planning

This chapter will introduce project planning and establish guidelines for what should be contained in a formal project plan.

Project Planning and Customer Needs

As was stressed in Chapter 1, the first order of business in today's world must be meeting the needs of customers (the "next-in-line"). If this is to be done in project management, the next-in-line must be identified and his or her needs defined.

This is often easier said than done. There is, first of all, identifying just who is the customer in the first place. Then there is defining needs. Often, customers have an "itch" that they want "scratched." That is the best definition they can offer. They want the product to be "easy to use." They want "convenience." These basic "itches" must be *translated* into product or service features. We can say that *solutions* are developed for customer needs.

One important point is that we want to do no more than necessary to satisfy customer needs, since that is wasting money. On the other hand, we do not want to do *less* than necessary, or we may lose the customer.

The contemporary approach to translating customer needs into product features is Quality Function Deployment (QFD). Since QFD is outside the scope of this handbook, the reader is referred to Dimancescu (1992) or Juran (1989), who explain how the process works.

The Project Notebook

In steps 1 through 8 of the model for managing projects, plans are being developed and ultimately signed off. This plan is usually housed in a loose-leaf notebook (or notebooks, in the case of very large projects). Subsequently, as the project is executed, progress reports, revisions, and so on will be placed in the notebook, so that when the job is finally closed out, the notebook provides a complete "track record" of the job from start to finish. This notebook is then placed in a central file so that anyone who needs to can refer to it as an aid in planning subsequent projects.

Following are the items that should be part of every project plan, and that should be in the notebook:

☞ A problem statement

☞ Project mission statement—formal for large projects, informal for smaller ones (see Chapter 5 for developing a formal mission statement)

☞ Project strategy, together with a SWOT (Strengths, Weaknesses, Opportunities, Threats) analysis supporting it

☞ Project objectives

☞ Documentation of QFD analysis or other means of translating customer needs into solutions

☞ Statement of project *scope*

☞ Contractual requirements: a list of all *deliverables*, including reports, hardware, software, and so on

☞ End-item specifications to be met, including building codes, government regulations, etc.

☞ Work Breakdown Structure

☞ Schedules: both milestone and working schedules should be provided

☞ Required resources, including people, equipment, materials, and facilities. These must be specified in conjunction with the schedule. Loading diagrams are helpful.

☞ Control System

☞ Major contributors. Use a Linear Responsibility Chart for this.

☞ Risk analysis with contingencies when available

The following are optional:

☞ Statements of Work (SOW)

Signoff of the Plan

Once the plan has been prepared, it should be submitted to *stakeholders* for their signatures.

> Stakeholder: Anyone who has a vested interest in the project. These include contributors, customers, managers, financial people, etc.

☞ A signature signifies that the individual is committed to his or her contribution, agrees with the scope of work to be done, accepts the specs as valid, etc. However, it is not considered a *guarantee*, since no one has 20/20 foresight, but it is considered a *commitment*, a promise to do everything within reason to meet project objectives.

☞ The plan should be signed in a *project plan review meeting*, not by mail!

☞ People should be encouraged to "shoot holes in the plan" during the review meeting, rather than waiting until problems develop later on.

Changing the Plan

It would be nice to think that a plan, once developed, would never change. However, that is unrealistic. Unforeseen problems are almost certain to arise. The important thing is to make changes in an orderly way, following a standard change-control procedure.

> Make changes in an orderly way, following a standard change procedure.

> The first rule of planning is to be prepared to replan!

If no change control is exercised, the project may wind up over budget, behind schedule, and hopelessly inadequate, with no warning until it is too late.

☞ Changes should be made only when a significant deviation occurs. A significant change will usually be specified in terms of percent tolerances relative to the original targets.

☞ Change control is necessary to protect *everyone* from the effects of scope creep.

☞ Causes of changes should be documented for reference in planning future projects.

Definition of Planning

Planning is the answering of the following questions:

☞ What must be done?

☞ How should it be done?

☞ Who will do it?

☞ By when must it be done?

☞ How much will it cost?

☞ How good does it have to be?

Suggestions for Effective Planning

1. Plan to plan. It is always difficult to get people together to develop a plan. The planning session itself should be planned or it may turn into a totally disorganized meeting of the type that plagues many organizations.

2. The people who must implement a plan should participate in preparing it.

3. The first rule of planning is to be prepared to replan. No one has 20/20 foresight. Unexpected obstacles will undoubtedly crop up.

4. Because unexpected obstacles will crop up, always conduct a
    risk analysis to anticipate the more likely ones. Develop a Plan B

just in case Plan A doesn't work. Why not just use Plan B in the first place? Because Plan A is better, but has a few weaknesses. Plan B has weaknesses also, but they must be different than those in Plan A, or there is no use in considering it a backup.

5. Begin with a definition of the purpose of doing whatever is to be done. Develop a problem statement. All actions in an organization should be taken to achieve a result, which is another way of saying, "solve a problem." If this step is skipped, you may find yourself developing the right solution to the wrong problem.

6. Use the Work Breakdown Structure to divide the work into smaller "chunks" that can have accurate estimates developed for duration, cost, and resource requirements.

Phased Planning

When a project spans a long period of time, or when considerable uncertainty exists about the approach to be taken (as in some research projects), it is impossible to plan far-term activities in much detail. The approach is to plan near-term work in detail, and as each phase is completed, to plan the next phase in detail.

While this is a valid approach, the politics of your organization may prohibit its application.

Project Planning Steps

The basic planning steps and the resulting documents that must be generated are as follows:

☞ Define the problem to be solved by the project. This yields a problem statement.

☞ Develop a mission statement (when appropriate), followed by statements of major objectives.

☞ Develop a project strategy.

☞ Write a scope statement to define project boundaries (what *will* and *will not* be done).

☞ Develop a Work Breakdown Structure (WBS).

☞ Using the WBS, estimate activity durations, resource requirements, and costs (as appropriate for your environment).

☞ Prepare the project master schedule and budget.

☞ Decide on the project organization structure—whether matrix or hierarchical.

☞ Set up the project notebook.

☞ Get the plan signed off by all project stakeholders.

Chapter 5

Planning: Developing Project Mission, Goals, and Objectives

Deciding What Must Be Done: Defining Your Mission, Goals, and Objectives

Importance of the Mission Statement

Without a clear understanding of its mission, a project team is like an airplane without a rudder. It will go wherever the

A mission statement provides the basis for which goals and objectives can be set and for making decisions, taking actions, hiring employees, etc.

winds blow, but not necessarily where it is intended to go.

As is true for an organization as a whole, a mission statement for a project gives it a sense of purpose and direction. It is a broad statement, from which all subsequent planning can proceed. It can be developed using a very formal procedure, which is presented later in this chapter, or it can be more informally stated. In the Gulf War, we were constantly reminded by the allied forces that their mission was to "free Kuwait." Nothing more. Nothing less.

The mission statement should be used to set goals and objectives, to make decisions, and to determine what goods and services the organization should be providing, whether it be a project group or overall company.

The Mission Identification Process

A mission statement should answer three questions:

1. What do we do?

2. For whom do we do it?

3. How do we go about it?

As an aid to answering these questions, it is useful for the team to work through the process outlined in the text box on page 49.

The last step in the sequence is to write the mission and purpose statement for the team. The procedure recommended for identifying the team's primary mission is shown on page 50.

Guidelines for Developing Sound Objectives

Before detailed planning can be carried out, specific objectives must be established. Those objectives *should be written out!* There are at least three reasons for this.

> ➤ Identify the team's *internal* and *external* environment.

> ➤ List all of the team's *stakeholders*.

> ➤ Highlight the team's *customers* from within the list of stakeholders just generated.

> ➤ Check the three most important stakeholders—at least one of them should be the team's major customer.

> ➤ Make a list of those things your three most important stakeholders want from the team.

> ➤ When the team has finished its job, how will members know they were successful?

> ➤ List those *criteria for success* that will be used to judge the team's performance.

> ➤ What critical events might occur in the future that could affect the team's success either positively or negatively.

> ➤ Now write the mission and purpose statement.

Developing a mission statement

Each individual prepares
a personal statement of
the team's primary
mission.

↓

Elements of the primary
mission that represent differences in priorities for
individual members are
identified so that they
can be managed.

↓

The group then combines
the individual views into
a team statement of the
primary mission.

↓

The group reviews and
critiques the meeting.

First, the discipline of writing out your objectives will force you to clarify them in your own mind. I have found that when I start to write out my objectives, I am sometimes not too clear on them myself.

Second, if they are in writing, everyone in the team has access to them and can refer to them periodically.

Third, being able to refer to them in written form should help members of the team resolve differences of opinion about what is supposed to be done.

Objectives may be to:

➤ **Develop expertise in some area**

➤ **Become competitive**

➤ **Improve productivity**

➤ **Improve quality**

➤ **Reduce costs**

➤ **Modify an existing facility**

➤ **Develop a new sales strategy**

➤ **Develop a new product**

In order to achieve clearly defined objectives, they should meet the following conditions:

☞ They should be **specific**—that is, not fuzzy, vague statements, such as "I want to be the best." What does that mean?

☞ They should be **measurable** when possible. This can be very difficult to achieve. How to you measure performance improvement of knowledge workers, for example?

☞ They must be **verifiable**—something you will be able to tell you have achieved. This is especially important when no measure exists for the objective. In that case, some sort of observable evidence must be available to ensure that the objective was actually met.

☞ Fit **higher level** organization objectives.

☞ Stated in terms of **deliverable items,** if possible. Deliverable items may be assessment reports, written recommendations, etc.

☞ They must be **comprehensible**—that is, *understandable*—stated in such a way that other people will know what you are trying to achieve. Have you ever left a meeting and wondered what everyone was supposed to do? Chances are the objectives were not stated clearly or in understandable language.

☞ They should be **time-limited** if possible. Remember this rule when setting performance-improvement objectives for employees. If such targets are not time-limited, they will never happen!

☞ Objectives should be **attainable.** That means that they should be both realistic and achievable. When appropriate, objectives should be **assigned a risk factor** so others in the organization will be aware of such risk.

☞ They should also specify a **single** end result. When multiple objectives are combined into one statement, it becomes difficult to sort out what is being said.

A Point About Definitions

I have found that people tend to confuse *tasks* and *objectives.* An objective is a desired end-state. You are presently at point A and want to get to point B. Tasks are those actions that you take to arrive at the final destination.

Determining what tasks or actions must be taken to reach an objective is part of problem solving and/or planning. Note that no statement of objective should specify *how* it will be achieved, since this may lock you into a method that is not the best to pursue. Keep the problem-solving process separate from setting objectives.

Establishing Priorities

In a project, many objectives will be sequenced strictly by *logical* considerations. However, others may have priorities that are a function of other factors of importance. Such importance may be determined by **need, economics,** or **social desirability.** Care must be exercised that

> Doing the right things is more important than doing things right.
> —Peter Drucker

less important objectives do not sidetrack progress toward more important ones.

Objectives that must be accomplished before some other target can be reached are called **feeder** objectives.

To prioritize your objectives, it may be enough to simply group them into categories A, B, and C—with "A" being most important, etc. On the other hand, you may need to actually rank-order your list. If you try to rank more than ten objectives, the task is very difficult. To make the job easier, you may want to use the method of paired comparisons.

Suppose I have four objectives that I want to rank-order. They are listed below.

1. Replace roof on house.

2. Enter the MBA program at local college.

3. Learn to play golf.

4. Take a trip to Europe.

Rather than trying to rank these by "brute force," I compare all possible pairs, and put a check beside the one in each pair that is most important to me. The number pairs are listed below:

1✓	2
1✓	3
1✓	4
2✓	3
2	4✓
3	4✓

As can be seen, objective one is the most important, with three votes, followed by objective four, two, and finally, three.

The Priority Matrix

Because there are so many comparisons to make for a large number of objectives, the method of paired comparisons can be greatly simplified using a priority matrix like the one shown in Figure 5.1. Consider six alternatives, which must be ranked. The matrix makes the comparison very straightforward. Following are the objectives to be ranked.

1. Install new grinder.

2. Develop standard test procedure for product x.

3. Recruit person for position y.

4. Do performance appraisal for Charlie.

5. Find second source for part z.

6. Review standards document for quality department.

Figure 5.1. Ranking Matrix.

	1	2	3	4	5	6	Total	Rank
1		1	0	0	1	1	3	3
2	0		0	0	0	0	0	6
3	1	1		0	1	1	4	2
4	1	1	1		1	1	5	1
5	0	1	0	0		1	2	4
6	0	1	0	0	0		1	5

The vertical axis is more important than the horizontal axis.

Chapter 6

Strategic Planning in Project Management[1]

Strategic Planning Defined

The word *strategy* comes originally from military terminology. The *Oxford English Dictionary*, considered to be the authoritative source of definitions, offers the following:

> *Strategy:* **the art of projecting and directing the larger military movements and operations of a campaign. The mode of executing tactics.**
>
> *Tactics:* **the art of handling forces in battle or in the immediate presence of the enemy.**

In colloquial terminology, we sometimes call strategy a "game plan," an overall approach to achieve our major objectives. The important point here is that strategy can only be decided upon once an organization's mission and objectives have been determined. All planning is done to meet objectives, and strategy is only the broad outline of the plan.

To give examples of strategy, we have JIT, concurrent or simultaneous engineering, and self-directed work teams as ways of achieving business objectives. The recent Gulf War also provides an example of military strategy. We learned from battles fought during World War II that one key to success (or failure) is *logistics*—the ability to supply troops with food, ammunition, and other supplies as well as to communicate with them. This understanding was employed in formulating allied strategy against Iraq: cut off their ability to supply and communicate with their troops, while simultaneously making sure we could supply our own. In fact, the bombing of bridges achieved two parts of that objective simultaneously. Not only could vehicles not carry supplies to Iraqi troops, but since telephone lines often ran under the bridges, destroying the bridges also cut the telephone lines.

Another example of strategy was just related to me recently. During World War II, Avondale Shipyards devised a new way of building ships. For centuries, shipbuilders have followed a basic approach, which is to build the ship in the position that it ultimately occupies when placed in the water. That is, the keel is on the ground and the decks are above.

Avondale decided to build the ship upside down. The reason—it was far easier to weld steel when it was below the welder than when it was above. Naturally, they had to devise a way of turning the ship over once the assembly was complete up to the decks, and they did. This strategy was so much more efficient than traditional approaches that Avondale had a significant competitive advantage over shipyards that employed the old method.

In developing project strategy, two fundamental questions must be answered by the project team:

☞ What are we going to do?

☞ How are we going to do it?

In the model for managing projects that was presented in Chapter 3, the first major step is to define the problem to be solved. That is, we must understand what the project is intended to do for the organization and the end user. Since the need to clearly define the problem is stressed in that chapter, it will not be emphasized here.

Once it is known what the project is supposed to achieve, specific objectives can be developed. The strategic planning phase must be a distinct part of the model. Timing is very important. If strategic planning is done too soon, it may be so vague (because of lack of sufficient definition) as to be useless. If it is too late, decisions may have already been made that will limit possible alternatives. In the model presented in Chapter 3, the step to develop alternative strategies for implementing the project follows the definition phase. To keep the overall model simple, strategic planning is shown as a single activity or step. This step, however, must be broken down into some substeps. In Figure 6.1 is a model to illustrate those substeps.

The Strategic Planning Model

The substeps in the Strategic Planning Model are not in any particular sequence. In fact, they are to some degree interactive, so that the information derived from one analysis might require going back to another step and digging out more information. Two of the components in the strategic planning model are contained in the model for developing a mission statement, presented in Chapter 5. These are the identification of the expectations of major inside and outside interests, which are called stakeholders in the mission-development model. This means that, if a mission statement has been developed

Figure 6.1. Strategic Planning Model.

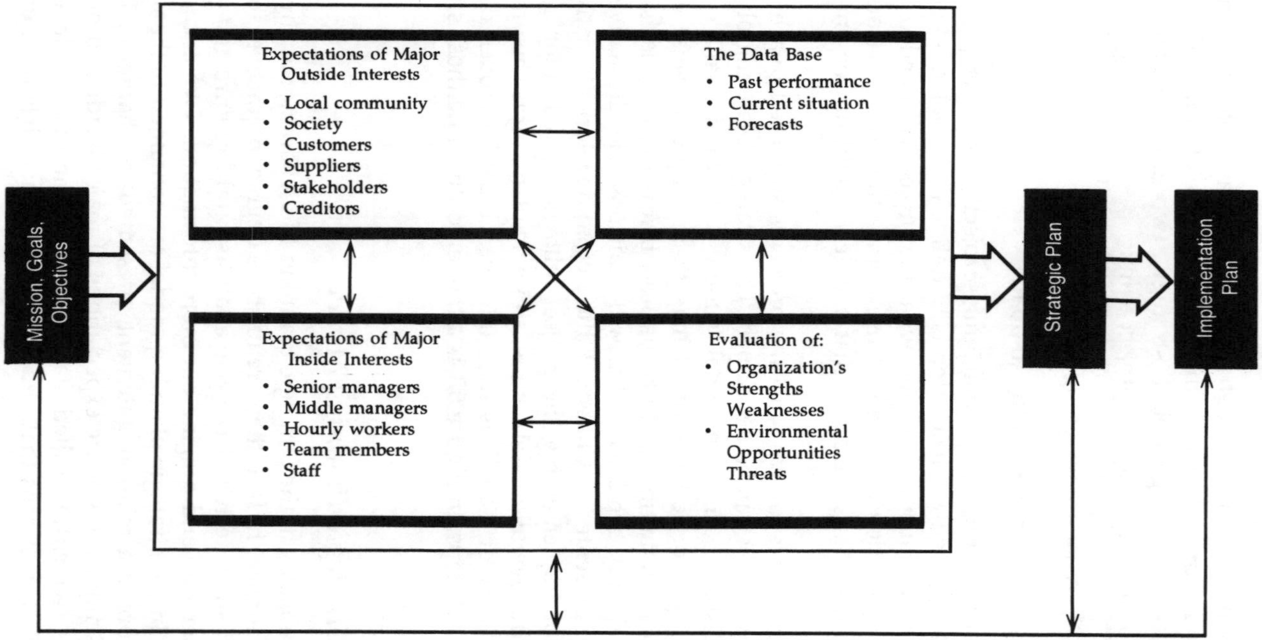

in accordance with the model, stakeholder expectations have already been identified. Also, for a project team, some of those stakeholders listed in the strategic planning model may be relevant while others may not. However, one should be careful not to dismiss a party as irrelevant without considering whether that conclusion is correct. Stakeholders are identified by whether the project may *impact them* in some way or whether they may have an impact on the project. If either is true, they should be considered important to the analysis.

Conducting the database analysis is limited in some organizations by a lack of good historical data. In that case, the analysis depends on the memories of individuals, and is subject to all of the biases and inaccuracies to which human beings are prone. Those limitations should be noted in making use of remembered data.

Analysis of the current situation should be easier, assuming that the project team members have access to vital information on the business, its competitors, and so on.

Forecasts are based on environmental scanning. They should include examination of technological developments, economic trends, pending government regulations, social trends, and so on. Naturally, forecasting is very difficult, and is limited by what information is available on competitors and other key entities.

Finally, we have the evaluation of environment and company for some specific variables. This evaluation is generally called a SWOT analysis. It prescribes that the team examine the company's *Strengths* and *Weaknesses* as well as the *Opportunities* and *Threats* presented by the environment.

These factors are clearly not independent. For example, forecasting is based on evaluations of the external environment and understanding the expectations of stakeholders. The current situation is also influenced by the environment and expectations of stakeholders. For that reason, the model is drawn to indicate that interdependence.

Conducting the Analyses

Environmental Factors

The major environmental factors that may affect a project are economic, technological, government or legal, geographic, and social. The economic variable can affect a project in many ways. During a recession, companies tend to run "leaner" than in more prosperous times, making resources more scarce and conflicts among projects almost inevitable. When the project spans national borders, currency fluctuations can be a significant factor. In addition, the economy in both the host country and the home country will be considerations. For those projects that span long time-frames, will inflation affect the project and, if so, what parts?

Technological changes can be the most difficult to forecast and deal with. In one of my own projects, for example, we wanted to design a new 1,000-watt linear amplifier with totally solid-state devices (no vacuum tubes). However, a feasibility study showed that current devices were not capable of meeting all of the technical requirements, so a conventional vacuum tube design had to be implemented.

E. F. Schumacher (1989) has observed that Westerners are inclined to do all projects with the highest technology available, when that might not be the best approach for projects in developing nations. He suggests that employing the right level of technology is important for the project to be judged a real success.

Projects are increasingly affected by government regulations and legal issues. Product liability suits in the United States have grown to such an extent that companies are very cautious in their handling of new products, construction, and so on. No one wants to make sports helmets, for example, because of the possibility of being sued if a player is injured. Environmental regulations have forced many businesses to change their way of running projects. At the NASA test facility near Las Cruces, New Mexico, for example, a test engineer told me that they once tested rocket components with minimal concern for the effluents. However, because of population growth near the test facility, they have had to take measures to contain toxic gases, so that no danger is posed to nearby residents.

Geographic factors certainly play a part in how some projects are run. Many companies are now global, and co-location of project participants is impossible. Fortunately, with modern communications technology, they are able to achieve what is called *virtual co-location*, whereby members of the team "meet" as often as necessary through teleconferencing. Naturally, geography also affects strategy in construction projects in terms of terrain, material resources available, and so on. The other influence is human resource availability. I recently saw the construction of a large Shell refinery in Bintulu, Sarawak. Ten years ago, Bintulu was a small fishing village of about 6,000. Now almost 60,000 people live there—naturally most of them imported.

It may be, of course, that key members of a project team will not want to spend long periods in very inhospitable locations, so this may require the recruitment and training of local personnel for the duration of the job.

Social factors are sometimes overlooked, especially by technical people, in planning project strategy. The social factor includes an assessment of the values, beliefs, traditions, and attitudes of people—in short, the culture of the people who are stakeholders in the project. Religious and other significant holidays must be factored into project scheduling. In January, people in the Far East celebrate Chinese new year much more than Westerners do their new year. Everything may come to a halt for a few days while people get ready and celebrate this important event.

By the same token, ignorance of the social *taboos* of a culture can create embarrassment and actual failures in projects. There are now books, tapes, and training programs aimed at helping people avoid cultural missteps, and I believe these will become increasingly important as the world becomes an even greater global economy.

In looking at all of the environmental factors, one asks, do they represent an *opportunity* or a *threat* to the success of the project? Technological developments, for example, can be either, depending on circumstances. If a design is frozen with a certain technology and a new development cannot be integrated with the design, then that change represents a threat to the success of the project, since acceptance by customers is likely to be low. Anyone designing conventional record players must certainly have been alerted by the development of CD players that there was a limited market for record players, and if they were really on their toes, they took steps to enter the CD market. Otherwise, they might have experienced the same decline that manufacturers of buggy whips did when the automobile displaced the need for their product.

Organizational Factors

Assessing an organization's strengths and weaknesses is a key element in strategic planning. Unfortunately, biases too often discredit the analysis. Managers are inclined to be optimistic about the strengths and a bit blind to the weaknesses of the organization. Nevertheless, the analysis must be done.

Factors to examine include expertise of personnel; labor relations; physical resources; experience with the kind of project being planned; company image; senior management attitudes; morale of employees; market position of the organization; tendencies to overdesign or miss target dates; commitment of the organization to supply resources to the project as promised. Naturally, you want to capitalize on strengths and minimize the impact of weaknesses. Further, those identified environmental threats must be capable of being offset by the team's strengths, and a conscious effort should be made to take advantage of opportunities presented by the environment.

Expectations of Stakeholders

Expectations of senior managers can be a major influence on the success or failure of a project. When those expectations for project performance are unrealistic, the impact is almost always negative. One of the more common expectations is that *all* target dates will be met. Such expectations lead to conflict.

The expectations of other stakeholders can also make or break a project. As an example, members of the community hear about the project and believe that it will create job opportunities for them. The project manager considers the skills needed for project success to be missing from the local community and recruits outsiders. There is public outrage, followed by unpleasant altercations, which results in senior management being pressured to abandon the project. And numerous examples exist of public pressures to abandon construction of hazardous waste facilities, nuclear power plants, and other projects considered by the community to be a threat to their security.

On the positive side, construction of the Saturn plant in Tennessee was undoubtedly aided by positive public reaction.

Formulating Project Strategy

Once all of these factors have been identified and examined, the project team is ready to develop a number of alternative methods of project implementation. These methods must meet external threats and take advantage of opportunities. Usually strategy will be a combination of several elements, as was the strategy in the Gulf War: (1) cut off Iraq's logistics capability; (2) enable our own logistics capability; (3) make Iraq think we were going to hit them from the sea, so they would divert defenses to the coast.

Coxon (1983) lists twelve possible strategies for projects:

1. Construction-oriented: an example of this would be the Avondale Shipyards approach to building ships.

2. Finance-based: this might involve some creative way of funding a project, perhaps through the use of bonds or grants; it might also involve special attention to cash flow and cost of capital.

3. Governmental: this involves taking into account government requirements and working closely with appropriate agencies to assure that no pitfalls will block progress.

4. Design: when certain design techniques have an advantage over others, this strategy may offer an advantage.

5. Client/contractor: this might include forming partnerships between client and contractor.

6. Technology: employing a cutting-edge technology might present certain risks, but offer greater competitive advantages. As mentioned previously, choosing the right level of technology might be important in developing countries.

7. Commissioning: if the commissioning aspects of the project are considered to be especially difficult or complex, then this strategy might be employed.

8. Cost, quality, or time: because these are interrelated, emphasizing one will be at the expense of impacting another. For example, when speed is of the utmost importance, and quality standards must be simultaneously maintained, then cost must increase. Nevertheless, if there is a significant market advantage to be gained through speed, the cost may well be offset by the profits made.

9. Resource: a resource strategy would be necessary when a particular resource is limited or abundant. For example, in Indonesia, Thailand, and other eastern countries, labor costs are so low that many construction projects are labor-intensive by western standards.

10. Size: it may be that for certain kinds of projects, economies of scale are only obtained once the size of the job exceeds a certain level.

11. Contingency: the strategy goes only so far as planning what to do if certain things happen.

12. Passive: this is a situation in which the project manager decides (consciously or unconsciously) to have no strategy at all (paradoxical, since this is in itself a strategy). It might be appropriate when the future is believed to be very stable or, conversely, to be so chaotic that developing a strategy is virtually impossible. This is also called "flying by the seat of the pants."

It is very important to conduct a risk analysis during project planning.

Conducting a SWOT Analysis

Following are the questions asked in conducting a SWOT analysis. The form on page 68 should also help simplify the process. The questions that must be answered are:

☞ What **Strengths** do we have? How can we take advantage of them?

☞ What **Weaknesses** do we have? How can we minimize the impact of these?

☞ What **O**pportunities are there? How can we capitalize on them?

☞ What **Threats** might prevent us from getting there? (Consider technical obstacles, competitive responses, values of people within your organization, and so on. Note that threats are not necessarily the same as *risks*.)

☞ For every obstacle identified, what can we do to overcome or get around it? (This helps you develop contingency plans.)

STATEMENT OF GOAL/OBJECTIVE _____

SWOT analysis- goals & objectives

LIST STRENGTHS	HOW CAN YOU BEST TAKE ADVANTAGE OF THESE?	LIST WEAKNESSES	HOW CAN YOU MINIMIZE THE IMPACT OF THESE?
WHAT OPPORTUNITIES DOES THIS PROJECT PRESENT?	HOW CAN YOU BEST TAKE ADVANTAGE OF THEM?	LIST THREATS: those risks or obstacles that might prevent success	HOW CAN YOU DEAL WITH EACH IDENTIFIED THREAT?

Implementation or Tactical Planning

Once appropriate strategy has been developed, an implementation plan must be produced that actually follows the strategy. The remaining chapters on planning deal with the implementation aspects of planning. See also the reprinted article by Slevin and Pinto (Chapter 14), which discusses the need to balance strategy and tactics and gives guidelines on how to do so.

Endnotes

[1] The content of this chapter was strongly influenced by an article by Robert Coxon, entitled "How Strategy Can Make Major Projects Prosper," *Management Today*, April, 1983. I have tried to present some of his ideas in a very general way, and have combined his thinking with my own experience in strategic planning.

Chapter 7

Using Work Breakdown Structures in Project Planning

Planning: Using the Work Breakdown Structure

In a previous chapter, I said that planning is answering some questions, among which were, "What must be done?" "How long will it take?" and "How much will it cost?" Planning the *what* is vital. The Work Breakdown Structure (WBS) provides a tool for planning the *what*, including estimates of resource requirements, activity durations, and costs.

The Case of the $600,000 Error

A project manager took over a project that was already in progress—the construction of a new wing of a hospital. The former project manager had left to take a new job. As he examined the plans for the project, the new manager felt uneasy. Something seemed wrong, but he couldn't find what it was. He told his boss, and got the same response. "Stay with it until you find what's wrong," said his boss.

A few days later, he found the problem—they were already doing site preparation with large earth-moving equipment, and it was nowhere in the plan. When he did an estimate, the site work alone was almost $600,000—on a job targeted originally to be around two million dollars! Imagine having to tell the board of directors of the hospital that they needed another $600,000 to complete the project.

Even though it seems inconceivable that anyone could omit the site work from a construction project plan, it is the omission of work that causes many projects to go over budget and miss their deadlines. Further, this often happens when the project manager plans the job all by himself, thereby violating one of the cardinal rules of planning, which is that the people who must implement a plan should participate in preparing it.

To avoid such problems, organizations should follow a standardized procedure for planning and managing projects. Such a standardized procedure was provided by the flow chart presented in Chapter 3. The flow chart has a signoff meeting at step 8, which in itself should aid in catching flaws in the plan. However, reviews are no substitute for participation in the planning process itself.

Estimating Time, Cost, and Resource Requirements

Every project manager is faced with the same problem—how do you estimate what it will take to do something? After you have done something once, estimating what it will take to do the next job is easier, but that does not mean that an exact determination of time, cost, and resources can be given. It will still be an *estimate,* and an estimate is *not exact!*

Estimating is never simple, and the higher the stakes, the more anxiety-provoking the job is. However, unless the estimating problem can be managed, projects will never come in on time or on budget.

The question is, how does a project manager know how long it will take to do a project, even given that he or she knows how many human resources are available to do the work? The answer is, *from experience!* That is the standard answer that everyone gives. However, it is not at all clear exactly what that means.

Let us examine what we mean by experience and its relationship to estimating activity durations. As an example, if you have been driving to the same workplace from the same home for several years, using the same route each time, you know about how long it takes to get to work. In large metropolitan areas, people report times like those shown in Table 7.1.

Table 7.1. Driving Times for Metropolitan Workers

Average time:	45 minutes
Shortest time:	35 minutes
Longest time:	60 minutes*

*Assuming no blizzards, etc.

When asked, "What is your best estimate of how long it will take you to get to work tomorrow?" they usually respond with, "forty-five minutes." That is, they give the average driving time.

However, if you ask, "How long would you allow yourself to get to work if you absolutely had to get there on time?" they say, "one hour." So they allow themselves the upper limit, just in case they are faced with a wreck or whatever causes their time to go to the upper limit. That way, they will be sure to get to work on time, unless the situation is extreme and the time exceeds the one-hour upper limit.

In a project, much the same process is involved. If a task has been performed a large number of times, the average duration is known, given a certain level of human resources to do the work, and this average can be used as the basis for an estimate.

There is only one difficulty. From statistics, we know that an average duration has only a 50/50 likelihood of occurring (see Figure 7.1). (The reader unfamiliar with statistics should consult a basic text, such as the one by Walpole (1974), cited in the reading list, to see why this is true.) That is, there is a 50 percent chance that it will take longer than the average duration and also a 50 percent chance that it could take less than the average. (There must be a "Murphy's Law" involved here: everyone knows that *no* work activity ever takes less than the average time!) Having a 50 percent chance of success in completing one's work is not very comforting, especially if the stakes are high and the organization is determined to have the project completed by some predetermined date.

It is for this reason that people are inclined to pad their estimates. As the normal distribution curve in Figure 7.1 shows, there

Figure 7.1. Normal Distribution Curve.

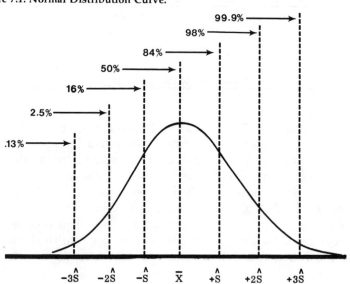

is an 84 percent probability that a duration one standard deviation above the mean can be met, 98 percent for two standard deviations, and 99.9 percent for three standard deviations. Thus, by padding, the project manager can greatly increase the probability that the work can be completed in the estimated time.

Unfortunately, safety carries a price. While increasing the allowed time increases the probability of a successful scheduled completion, it also increases the budgeted cost of the project, and it is possible that such an increase will increase the total estimated project costs to the point that the job will most likely not be funded.

For this reason, one of the assumptions made in project estimating is that *average durations are used, unless specified otherwise.* The idea is that, for a project consisting of a large number of activities, some of the work will take longer than the average estimated duration, while other tasks will take less than the average, so that the total project completion time will gravitate toward the mean expected time for the critical path! Again, Murphy's Law seems to prevent such an occurrence.

While padding is certainly justified to reduce risk, my opinion is that it must be done above-board, on a task-by-task basis. Otherwise, the project manager might assume that every member of the team has provided average-duration estimates and then put some padding into the project at the top level, thus adding "fat" to that which individuals have already put into their estimates, and then the project is sure to be too expensive.

This is one reason for not getting into game-playing in an organization. It sometimes happens that a manager asks for an estimate, and when it is provided, the manager cuts the estimate by 10 percent or 15 percent, based on the belief that the estimate contains at least that much "fat." Note that the probability of achieving a time one standard deviation below the mean is only 16 percent, so cutting an estimate that was an average expected duration severely reduces the probability that the time can be met.

If the project manager gets burned because his estimates were averages and they were cut, then the next time he is asked for an estimate, he will indeed pad, since he expects that his estimate will be cut. This time, however, the manger cuts 20 percent, so the next time the project manager pads 25 percent, and so on. Such "games" are hardly productive.

The objective of all project planning should be to develop a plan that is *realistic,* so that managers can make decisions as to whether to do the work. The objective should not be to try to "get" the project manager when he is unable to meet unrealistic deadlines.

It is instructive to consider the cause of variations in working times, using the driving example as a guide. Suppose we agree that the time it takes to drive to work varies widely because of unpredictable traffic-flow patterns. If we could eliminate those, then the variation in driving time could be greatly reduced.

Indeed, suppose there was absolutely no other traffic on the road besides the single driver. (Perhaps this can be accomplished by leaving very early in the morning, before other drivers get out on the road.) In that case, there should be no variation in driving time—right? Of course that is not true. Even though the variation would be reduced, no driver can maintain an exactly constant speed, or even the exact-same variation at the same point in the road day after day.

For this reason, there will always be some variation in working times, caused by *factors outside the control* of the operator! It is therefore unrealistic to expect that anyone can estimate precisely how long a task will take to complete. (As someone has said, anyone who has 20/20 precision in forecasting should leave project management and invest in the stock market. The returns would be far greater than anyone could make managing projects.)

Other Factors in Estimating

An average expected duration for an activity can only be developed by assuming that the work will be nearly identical to work previously done and that the person(s) being assigned to the task will have a certain skill level. If a less-skilled person is assigned to the work, it can be expected to take longer, and conversely, a more-skilled person could probably do the job faster. Thus, adjustments must be made for the experience or skill of the person(s) assigned to the task.

However, we also know that there is no direct correlation between experience and speed of doing work. It may not actually be true that the more experienced person can do the job faster than the person with less experience. And putting pressure on the individual will pay dividends only up to a point. I remember once when I was pressuring one of my project team members to get a job done faster, and he got fed up with the pressure. Finally, he said, "Putting two jockeys on one horse won't make him run any faster." And he was right.

Then there was a project manager who told me that when his boss doesn't like an estimate of how long it will take to perform a task, he tells him to *use a more productive person!* That was his solution to all scheduling problems (Figure 7.2).

Figure 7.2. More Productive Person.

Another factor is how much productive time the person will apply to the task each day. It is not uncommon in some organizations for people to spend an average of 25 percent of each day in meetings, on the phone, waiting for supplies, and other activities, all of which reduce the time available to spend on project work. Allowances must be made for such non-project time.

In addition, if experience with a task is minimal, the expected duration might be adjusted upward, compared to the closest task for which experience exists. If virtually no experience can be used as a basis for estimating, then it might be appropriate to use PERT techniques (see Chapter 11). Another possibility is to use DELPHI or some other estimating method. For an application of DELPHI method to project estimating, consult the book by Burrill and Ellsworth, cited in the reference list.

For construction projects, there are books available containing *means tables*, which list the average expected durations for typical construction activities, together with "fudge factors" to be used in adjusting those times to compensate for geographic location, weather, and so on. A source of means tables is listed in Chapter 22.

For other types of projects, unfortunately, there are no means tables available, so historical data must be developed by keeping records on previous project work. This is, perhaps, one of the most important benefits of developing a standardized project management methodology—by doing the work in specified ways and by keeping records on actual working times, an organization can develop a database that can be used to greatly improve future project estimates.

Work Breakdown Structure Format

There are two popular forms of the Work Breakdown Structure (WBS). One looks similar to an organization chart, and might even be thought of as such, except that the boxes represent work activities, rather than reporting structures. The second form of WBS is the line-indented form. It is a straightforward listing of project activities, with each new indentation being a lower level of detail (smaller unit of work to be performed).

A commonly used format for WBS has six levels, named as shown in Figure 7.3. It is perfectly acceptable to use more than six levels, but you will have to devise names of your own for the lower levels.

An example of a WBS is also shown in Figure 7.4. This is a simple illustration of a project to write a book.

Figure 7.3. WBS Level Names.

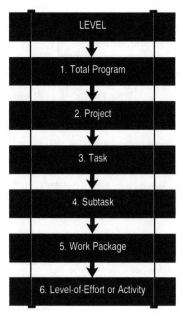

In Figure 7.5 is an illustration of estimates being made at the work package level. The level six activities can be estimated for resources, time, and labor costs.

In Figure 7.6 is a line-indented form of the WBS, previously shown in Figure 7.4. This is a convenient format to use, in that it can be produced entirely in text format on a computer, complete with line numbering. However, it does not capture the visual scope of the project as well as the graphic form.

General Aspects of Work Breakdown Structures

☞ Up to 20 levels can be used. More than 20 is considered overkill.

Figure 7.4. Sample Work Breakdown Structure (WBS).

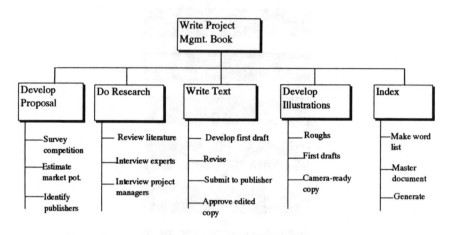

Note: This WBS begins at the project level and goes down to the subtask level. A large project might go on down to level six or beyond.

An Example of a Work Breakdown Structure

Figure 7.5. Estimate at Work Package Level.

Example of a Work Package Cost Estimate

Develop Illustrations for Book

Task	Time Req'd	Labor Rate	Total Cost
Pull clipart	24 hrs.	10.00	240.00
Computer art	40 hrs.	15.00	600.00
Pasteups	100 hrs	10.00	1000.00
Type for art	10 hrs.	10.00	100.00
Make stats	20 hrs	12.00	240.00

Work Package Cost $2180.00

Figure 7.6. Line-Indented WBS.

Write Project Management Book
1. Develop Proposal
   1.1 Survey competition
   1.2 Estimate market potential
   1.3 Identify publishers
2. Do Research
   2.1 Review literature
   2.2 Interview experts
   2.3 Interview project managers
3. Write Text
   3.1 Develop first draft
   3.2 Revise
   3.3 Submit to publisher
   3.4 Approve edited copy
4. Develop Illustrations
   4.1 Roughs
   4.2 Final drafts
   4.3 Camera-ready copy
5. Index
   5.1 Make word list
   5.2 Master document
   5.3 Generate

☞ All paths on a WBS do not have to go down to the same level.

☞ The WBS does not show sequencing of work except in the sense that all level-five work packages hanging below a given subtask must be complete for the subtask to be complete, and so on. However, work packages below that subtask might be performed in series or parallel. Sequencing is determined when schedules are developed.

> A work breakdown structure *does not show the sequence in which work is performed!* Such sequencing is determined when a schedule is developed.

☞ A WBS should be developed before scheduling and resource allocation are done.

☞ The WBS should be developed by individuals knowledgeable about the work. This means that levels will be developed by various groups and then the separate parts combined.

☞ Break down a project only to a level sufficient to produce an estimate of the required accuracy.

Using Charts-of-Accounts in Project Tracking

Ultimately, it will be necessary to compare actual progress on the project to the plan. In particular, labor costs will be charged back to the project, and accomplishment of work will be compared to the plan. The device for tracking costs is the chart-of-accounts. A sample is shown in Figure 7.7.

The Linear Responsibility Chart

A common problem in organizations is trying to determine who has responsibility for which tasks. The standard organizational chart is of the pyramidal variety. It portrays the organization as it is "supposed" to exist at a given point in time. However, the pyramidal organization chart is insufficient because it does not display the nonvertical relations between members in the organization. Although not normally defined, the interaction between people in a working environment affects the success of the effort and cannot be overlooked.

One of the most common problems of interaction in a project seems to be that someone makes a unilateral decision about something that actually affects one or more other individuals in the project. For example, capital equipment may be purchased without consulting other users to determine their needs, and the purchased equipment may be lacking. Or in designing a product, one designer

Figure 7.7. Chart of Accounts.

Account Number	Activity Description	Account Number	Activity Description
000 *		032	Camera work
001	Development of concepts	033	Office layout
002	Preliminary design	034	Reserved
003	Computer analysis	035	Reserved
004	Environmental tests	036	Contract administration
005	Alternative selection	037	Contractor payroll certifica-
006	Delphi technique		tion
007	Systems analysis	038	Reserved
008	Reserved	039	Reserved
009	Field investigation	070 *	
010 *		071	Project management
011	Final design	072	Project planning & schedul-
012	Draft specifications		ing
013	Drafting/graphics	073	Project coordination
014	Checking drawings	074	Reserved
015	Specifications review	075	Client meetings & confer-
016	Maintenance work		ences
017	Technical writing	076	Public meetings & hearings
018	Cost estimating	077	Reserved
019	Bid preparation	078	Reserved
020 *		079	Project review meetings
021	Quality control checks	080 *	
022	Reserved	081	Administrative services
023	Reserved	082	Clerical support
024	Computer data preparation	083	Composing & editing
025	Computer analysis	084	Typing
026	Computer keypunching	085	Reproductions/printing
027	Reserved	086	Training
028	Shop drawing review	087	Marketing & sales
029	Reserved	088	Reserved
030 *		089	Reserved
031	Prepare visual aids	090 *	

does her work without consulting another designer about aspects of the design that concern both of them.

Linear Responsibility Charts (LRCs) help by showing such requirements as *must be consulted* or *must be notified.*

Figure 7.8 is an example of a filled-in Linear Responsibility Chart. The empty form shown in Figure 7.9 can be copied and used in your own projects.

Figure 7.8. Linear Responsibility Chart (sample).

Project: Notebook for Proj. Mgrs. Date Issued: 01-Dec-90 Sheet 1 of 1
Manager Jim Lewis Date Revised: 13-Dec-92 Filename: LRCSAMP

Project Contributors

Task Descriptions	Len	Ann	Susi	Jim	Norm S.	Carolyn
Design Forms			2	1		
Final layout of forms		2	1			
Write guidelines for use				1	2	
Design package			1	2	2	
Develop sales plan				1	2	2
Production coordination		2	1	2	2	

Codes: 1 = Actual Responsibility || 2 = Support || 3 = Must be notified || Blank = Not involved

Figure 7.9. Linear Responsibility Chart.

Project: _____   Date Issued: _____   Sheet _____ of _____

Manager: _____   Date Revised: _____   Filename: _____   LRCFORM

Task Descriptions	Project Contributors									

Codes: 1 = Actual Responsibility || 2 = Support || 3 = Must be notified || Blank = Not involved

(c) 1991 by James P. Lewis

Forms for Planning

As an aid to the planning process, the forms on the following four pages are provided. They can be copied by the original purchaser of this book without additional permission from the publisher.

Figure 7.10. A Project Planning Form.

Project Action Plan			
Project:			
Prepared by:		**Date:**	
Objective:			
Measure of Accomplishment:			
Steps to take	Time Required	Resources Needed	Who Does It?

File: QPRODATA\FORMACT

Figure 7.11. Project Resource Planning Form.

Project Resource Plan				
Project:				
Prepared by:		Date:		
Resources	Person to Contact	How Many/ Much Needed	When Needed	Check if Avail.
People:				
Materials:				
Equipment, Tools, Facilities, Space:				
Special Services:				
Typing/Clerical: Reproduction: Others:				

File: QPRODATA\FORMRES

Figure 7.12. Form to Help Sell a Project Plan.

Selling The Project Plan	
Project:	
Prepared by:	Date:
People I need to convince of its value:	
What do I need from these people? o Authorization o Assignment of personnel o Equipment purchase o Support	
Advantages of accepting the plan	How can I best present this advantage?
Possible objections to the plan	How can I overcome these objections?

File: QPRODATA\FORMSELL

Figure 7.13. Risk Analysis, Contingency Form.

Contingency Planning	⌁
Project:	
Prepared by:	Date:
Even with the best of plans, things will go wrong. By conducting "what-if" analyses, the impact of such problems can be minimized. List below some problems that could adversely affect your project and formulate some possible solutions.	
Potential Problem	Solution Options
Notes:	

File: QPRODATA\FORMPROB

Section Three

Scheduling with CPM and PERT

Chapter 8

Developing the Project Schedule

A Brief History of Scheduling Techniques

> We can't leave the haphazard to chance!
> —N. F. Simpson

The earliest scheduling method worked out in detail was the bar chart, developed by Henry Gantt, and subsequently called the "Gantt chart." As Figure 8.1 on the next page shows, bar charts are very easy to construct and easy to read, with one exception. By looking at the chart, it is impossible to tell whether tasks *analyze* and *startup* plan are constrained to start immediately after *recommenda-*

Figure 8.1. Bar Chart for a Small Project.

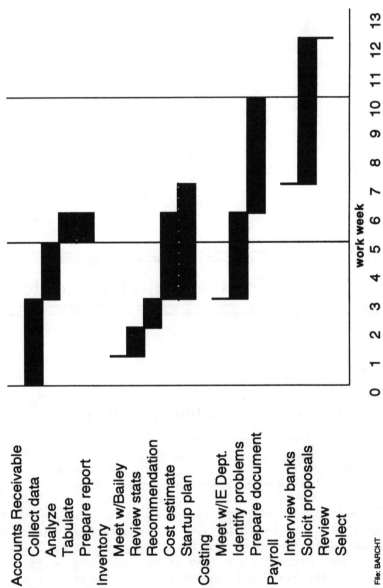

File: BARCHT

tion is completed, or if there is some other reason why they just happen to start at the same point when *recommendation* ends. Unless this information is available, one cannot tell what happens to a project if some activity gets behind (*slips,* in common terminology).

☞ A primary weakness of the bar chart is that it does not show the **interrelationships** between the various tasks being done.

☞ Trying to construct a schedule using the bar chart is likely to yield an unworkable arrangement, since the interrelationships may be missed.

☞ During the 1960s, the Navy, in conjunction with the Booz, Allen & Hamilton Inc. consulting group, developed a scheduling system called PERT (**P**erformance **E**valuation and **R**eview **T**echnique), which was applied to the Polaris project. By their estimates the project was shortened about two years through the use of the PERT method.

> **P** = **performance**
> **E** = **evaluation &**
> **R** = **review**
> **T** = **technique**

☞ About the same time, Du Pont developed CPM (**C**ritical **P**ath **M**ethod).

☞ CPM is very similar to PERT. The major difference between PERT and CPM is that PERT applies statistics to networks, whereas CPM does not.

> **C** = **critical**
> **P** = **path**
> **M** = **method**

☞ It is technically incorrect to talk about CPM/PERT as if they are the same, although common usage has led to that terminology.

Definitions of Network Terms

ACTIVITY

An activity always consumes time and may also consume resources. Examples include paperwork, labor negotiations, machinery operations, and lead times for purchased parts or equipment.

> Using WBS (Work Breakdown Structure) terminology, an activity could be a work package or a level-of-effort. Occasionally, an activity might even be drawn at a subtask or task level.

CRITICAL

The word *critical* is applied to any activity or event that must be achieved by a certain time and has no latitude whatsoever. Failure to achieve the specified time will result in the project end date slipping.

CRITICAL PATH

The *critical path* is initially the longest path through a project network and determines the earliest date on which work can be completed. It is also generally set up to have no latitude (called *float*).

DUMMY

An arrow that denotes nothing but a dependency of one activity upon another is a dummy activity, and carries a zero duration. Dummies are usually represented by dashed-line arrows.

EVENT

Beginning and ending points of activities are known as events. An event is a specific point in time. Events are commonly denoted graphically by a circle, and may carry identity nomenclature. (Words, numbers, alpha-numeric codes, etc.) Note that an event is *binary;* that is, it is either achieved or not, whereas an activity can be partially complete.

ij NOTATION   The numbering of events in an arrow diagram. The *i-number* is always the beginning event, while the *j-number* is always the ending event.

MILESTONE   An event that represents a point in a project of special significance. Usually it will be the completion of a major phase of the work, and often a project review or audit will be conducted when the milestone is achieved.

NETWORK   Networks are also called arrow diagrams. They provide a graphical representation of a project plan, showing the relationships of the activities.

Network Techniques in Project Management

Both CPM and PERT make use of arrow, or network, diagrams to show the interrelationships of work in a project. Several different graphical schemes for drawing networks exist. The two leading schemes are called *activity-on-arrow* and *activity-on-node.*

While it is possible to do network computations manually, it is not very practical. For small networks, it is easy to compute working times, but even then, converting these to calendar dates becomes a bit cumbersome. For another thing, software to run on personal computers (PCs) has become so inexpensive compared to the cost for labor that almost anyone can justify the expense. For that reason, your choice of notational systems will be largely determined by the software chosen. And, although there is software available that does activity-on-arrow notation, most software for personal computer applications does activity-on-node notation.

Network Examples

Figures 8.2 and 8.3 are examples of networks drawn using both activity-on-node and activity-on-arrow notation.

Both networks are drawn showing the same sequence of activities. Activity A is done before activity B, but activity C can be done in parallel with them. All of them must be completed before activity D can be started.

Figure 8.2. Activity-on-Arrow Network.

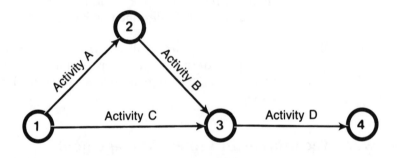

Figure 8.3. Activity-on-Node Network.

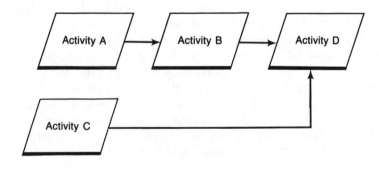

For the activity-on-arrow diagram, the activities can also be referred to by their event numbers. Thus, activity A can also be called activity 1-2; activity B is activity 2-3, and so on. Another common way of doing this is to let the *ij* numbers be subscripts. Thus, we might use the letter A for activity and have A_{ij}. For activity A, we would then write $A_{1,2}$. Note that the comma is used because an activity connected between events 10 and 11 without a comma would be confusing: A_{1011}.

In activity-on-node notation, events are not shown for each activity. They are *implied*, of course. Some people say that activity-on-arrow notation is *event-oriented*, while activity-on-node notation is *activity-oriented*. It is possible to use a node to represent an event in activity-on-node notation. These will generally be *milestones*.

Overlapping Work—Ladder Networks

In the diagrams shown so far, all of the work is shown being wholly sequential or wholly parallel with some other activity. Naturally, that is not always necessary. It may be possible to perform a small portion of a task, then start the next one, rather than waiting until the first task has been completed. When work is performed in this manner, we say that tasks are *overlapped*.

One factor that may require the overlapping of work is that, in most cases, both the start and finish dates for a project have been "given" before the project manager sits down to plan the work. The typical scenario is that the network is developed, the end date computed, and it won't "fit" between the two given dates. The critical path is too long, and must be shortened. (It is the old problem of trying to fit 10 pounds of trash in a five-pound bag.)

It may be possible to shorten the critical path by putting more people on the task, although all work has a point of diminishing returns.

Another alternative is to overlap the work. Consider the following example. Suppose you want to build an underground pipeline that will be several miles long. Once the pipe is laid, you will fill in the trench. Clearly, however, to wait until the entire length of pipe is laid before beginning to fill the trench leads to an unnecessarily long schedule. It should actually be possible to begin filling after a few lengths of pipe have been laid. This is diagrammed using the ladder network, as shown in Figure 8.4.

Figure 8.4. Ladder Network.

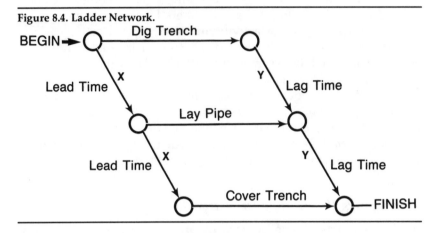

The delay before the subsequent activity (next section of pipe) can start is called a "lead" element, and is designated by an "x" in the figure. The "y" elements represent the amount of lag time that must pass from the completion of the digging operation to the completion of the last bit of pipe that is laid. We now have a path from beginning to ending that is shorter than that obtained by simply tying the three activities in series. One of the approaches to shortening schedules is to look for ways to overlap work.

In order to understand the times for the lead and lag elements in a ladder network, the reasoning is as follows. Consider the digging operation. You must dig a length of trench sufficient to hold at least a couple lengths of pipe before it makes sense to start laying pipe. In fact, you may elect to dig ten pipe-lengths of trench before you begin laying pipe. Therefore, the amount of lead time before pipe laying begins will be the time required to dig the desired length of trench. Similarly, before you begin filling in the dirt, you need to lay several lengths of pipe, connect them, and check the joints. Therefore, the lead time before filling will be the amount of time required to lay the pipe, connect it, etc.

For the lag elements, consider the completion of the digging. From the moment when digging the trench has been completed, until the pipe laying has been completed, will be the amount of time required to lay the remaining sections of pipe. If it takes ten hours to lay the last bit of pipe, then the lag from completion of digging until completion of pipe laying will be ten hours. The same is true of the lag from completion of pipe laying to completion of the filling operation. The lag will be the time required to put the last bit of dirt in the ground.

Note that some software will not properly analyze ladder networks, since the analysis must treat the entire ladder as a contiguous "chunk," rather than as separate elements. If this is not done, you may wind up with slack time in your lead and lag paths, which is not what you want.

Precedence Diagramming

When activity-on-node notation is used, the ladder network can be handled with precedence diagramming, which is an extension of the basic node method. It handles various logic relationships involving lead or lag elements using simpler notation than would be possible with conventional arrow notation. Networks are shown in Figures 8.5 through 8.9.

Finish-to-Start Relationship

Start of B must lag five days after the finish of A. A good example of using a finish-to-start element would be that activity A is the pouring of concrete, which must cure for five days before the next step (activity B) can be taken. Similarly, waiting for paint to dry or epoxy to cure could be diagrammed in this fashion.

Figure 8.5. Finish-to-Start Network.

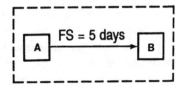

Start-to-Start Relationship

Start of B must lag three days after the start of A. This is the lead element in a ladder network. The digging starts. A few hours or days later the pipe laying starts.

Figure 8.6. Start-to-Start Network.

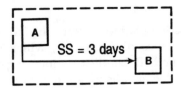

Finish-to-Finish Relationship

Finish of B must lag four days after the finish of A. This is the lag portion of the ladder network. The digging finishes, then four days later the pipe laying is completed.

Figure 8.7. Finish-to-Finish Network.

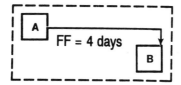

Start-to-Finish Relationship

Finish of B must lag 35 days after the start of A. There are not many practical examples of this logic being used, although it is possible.

Figure 8.8. Start-to-Finish Network.

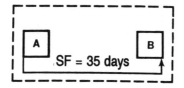

Composite Start-to-Start and Finish-to-Finish

Start of B must lag four days after the start of A and the finish of B must lag four days after the finish of A. This is the conventional ladder or lead-lag network.

Figure 8.9. A Composite Start-to-Start and Finish-to-Finish Network.

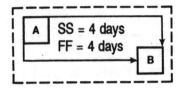

Common Pitfalls to Avoid

Following are examples of some of the more common mistakes made in using networks.

☞ **Subjective Network:** Work is diagramed in series when it could actually be done in parallel.

Figure 8.10. Lawn Project with all Activities in Series.

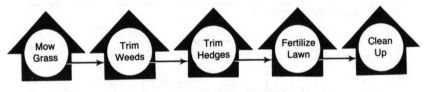

Figure 8.11. Lawn Project with Parallel Activities.

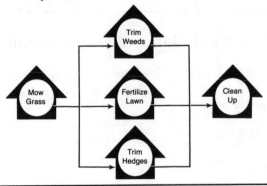

☞ **Partial Dependencies:** One activity is shown as being dependent on the total completion of another, when it could be started following completion of only a part of the predecessor. The general term for this is called *activity splitting* or *overlapping work*. (Figure 8.12).

☞ **Loops:** A loop is drawn in the network. Computers don't like loops, and always test for them before performing network computations. If a loop is found, the software will tell you and no calculations will be done until the loop is removed. (Figure 8.13).

Figure 8.12. Partial Dependency Network.

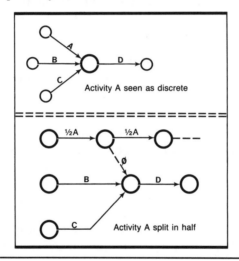

Activity A seen as discrete

Activity A split in half

Standard operating procedure: The networks should be sketched on paper before data is entered into the computer. This will help you avoid loops and will enable you to specify predecessor-successor relationships as you enter data. You can also mark off activities as they are entered into the computer, so that you can be sure nothing is forgotten.

Figure 8.13. Loop in Network.

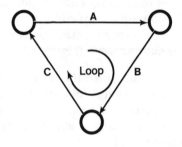

☞ **Natural and Resource Dependencies:** A natural dependency is one that is based on nothing but the logical way in which work can be done. A resource dependency is based on the availability of a person, facility, etc.

> Networks should be drawn based on *natural dependencies* only, and resource limitations should be handled afterward.

☞ **Fixed-Time Activities:** Some activities require fixed time periods outside the control of management. Examples: legal minimums (30 to 90 days advance notice for public hearings), technical minimums, such as curing time for concrete, etc. When these become part of the critical path, it is useful to try to accomplish in parallel other activities that can be done. However, when possible, it is advisable to keep fixed-time activities off the critical path.

☞ **Accounting for the Weather:** One method is to estimate how long a project will take, then to add to that duration a fudge-factor based on what part of the year the project will be done.

A better method is to consider each activity individually and see what impact weather might have on that activity specifically. It may also be possible to schedule activities that have large amounts of float during those periods when weather is likely to cause delays.

Level of Detail

The basic rule-of-thumb is that the network should be planned in no more detail than can be managed. For example, a WBS may be developed down to level six, but the schedule may show only work packages (level five). This is *not* a suggestion to play games with others in the organization. Rather, it is an acknowledgment that we cannot manage with precision at too great a level of detail.

Other questions to consider are:

1. Who will use the network, and what are their interests and span of control?

> Don't plan in more detail than you can manage!

2. Is it *feasible* to expand an activity into more detail?

3. Will accuracy of logic or time estimates be affected?

4. Do some skills, facilities, or other areas require more detail?

Conventional Assumptions

Three general assumptions are usually made in project scheduling. As is always true of assumptions, these must be understood to avoid problems.

1. The time estimate made for an activity is the *mean* or *average* time that it should take, and the estimate is called the *activity duration*. Estimates of activity duration do not include uncontrollable contingencies such as fires, strikes, or legal delays.

> If this assumption is violated by padding estimates, it should be done *above board*.

Don't plan in more
detail than you can control!

2. In estimating activity duration, the activity should be considered independently of those that precede or follow it. For example, if an activity depends on the delivery of some materials, the delivery should be shown as an activity with its own duration. It should not be said that the second activity will probably be late because delivery of materials may not happen on schedule.

> The assumption of independence is less likely to cause problems than the other two.

3. It is also usually assumed that a normal level of personnel, equipment, and other resources will be available for each activ-

ity. Since activities will inevitably run in parallel, which make use of the same resource(s) this will lead to an unmanageable schedule. Note that the assumption is equivalent to saying that you have *unlimited resources,* which is never the case. As was previously stated, however, initial scheduling is done using only logical dependencies and resource allocation is handled later.

> This assumption is *guaranteed* to be a killer, unless resource allocation methods are applied to the schedule.

Chapter 9

Scheduling Computations in Critical Path Method

Critical Path Method

Once a suitable network has been drawn, with durations assigned to all activities, it is necessary to determine where the longest path is in the network, and see if it will meet the target completion date. Since the longest path through the project determines minimum project duration, if any activity on that path takes longer than planned, the end date will slip accordingly, so that path is called the *critical path*.

> The *critical path* is the longest path through a project
> network, and thus determines the earliest completion
> for the work.

In the simplest form, computations are made for the network assuming that activity durations are exactly as specified. However, activity durations are a function of the level of resources applied to the work, and if that level is not actually available when it comes time to do the work, then the scheduled dates for the task cannot be met. It is for this reason that network computations must ultimately be made with resource limitations in mind. Another way to say this is that *resource allocation* is necessary to determine what kind of schedule is actually *achievable!* Failure to consider resources almost always leads to a schedule that cannot be met.

> Failure to consider resource allocation in scheduling
> almost always leads to a schedule that cannot be met.

Still, the first step in network computations is to determine where the critical path is in the schedule and what kind of latitude is available for noncritical work, under *ideal conditions*. Naturally, the ideal situation is one in which unlimited resources are available, so the first computations made for the network are done ignoring resource requirements. It is that method which will be described in this chapter, and resource allocation methods will be presented in the next chapter.

> Initial schedule computations are made assuming that
> unlimited resources are available. This yields the *best-
> case solution.*

Network Rules

In order to compute network start and finish times, only two rules are applied to *all* networks. These are listed on the next page as rules 1 and 2.

Other rules are sometimes applied by the scheduling software itself. Some of the more common ones are listed as rules 3 through 6. These are strictly a function of the software and are not applied to all networks.

Rule 1:   Before a task can begin, all tasks preceding it must be completed.

> Rules 1 and 2 apply to all networks.

Rule 2:   Arrows denote logical precedence. Neither the length of the arrow nor its angular direction have any significance. (It is not a vector, but a scalar.)

Some of the more common rules imposed by software follow:

Rule 3:   Event numbers cannot be duplicated in a network. This rule applies only to activity-on-arrow networks, and can be violated if the project is broken into subprojects.

> Rules 3–6 apply only if the software says so. They are *not* *universal!*

Thus, you could have an event 2 in subproject A and an event 2 in subproject B.

Rule 4:   Any two events may not be directly connected by more than one activity. There is no inherent reason for this—it is simply that the programmer is keeping track of the activities by the event numbers. See Figures 9.1 and 9.2 for an illustration.

Rule 5:   Networks are allowed only one initial event (nothing preceding) and only one terminal event. See Figure 9.3 for details.

Figures 9.1. Parallel Activities.

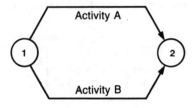

Figures 9.2. Parallel Activities Rule 4.

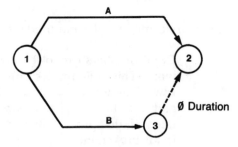

Figures 9.3. Project with Two Beginning Events.

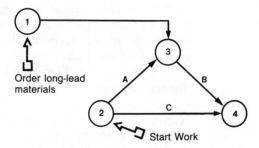

Rule 6:   Event numbers must flow from left to right in ascending order. This is a particularly onerous rule, since it requires that following events be renumbered if a new activity is inserted into the network.

Basic Scheduling Computations

Scheduling computations will be illustrated for activity-on-arrow notation, using the network shown in Figure 9.4. For simplicity, the arrows have been given letter names. However, they can also be referred to by their *ij* numbers, as was mentioned in the previous chapter. The network shown could be any kind of project, and we will not be concerned in this chapter about the nature of the work or resource allocation.

Figure 9.4. Network to Illustrate Computations.

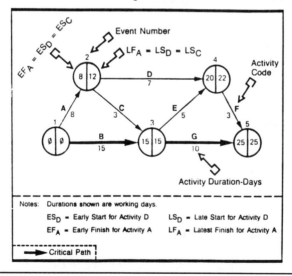

For this network, we will assume that the project starts at some *time = zero*, and we will determine how many working days are required to complete the job. Calendar dates could then be found by placing the starting point on the appropriate date, then counting forward the number of working days determined by the calculations, keeping in mind that nonworking days are not counted. These would include weekends, holidays, and perhaps vacation days. Naturally, scheduling software will do calendar determinations for you, but in this chapter, we will not be concerned with actual calendar dates. These will be illustrated in the next chapter.

Notation Used in the Computations

The following notation will be used in performing all computations:

EF   Early finish. When a subscript number (EF_1) is used, it will designate the early finish for an event. If a subscript letter (EF_A) is used, it will mean the earliest finish for an activity. Note that we could also refer to an activity by its *ij* numbers, so we could have $EF_{1,2}$, meaning the Earliest Finish for Activity 1-2.

LF   Late finish. Again, subscript number means event, while subscript letter means the activity.

ES   Earliest start for an activity.

LS   Latest start for an activity.

D_{ij} The duration of an activity.

Forward-Pass Computations

A forward pass is made through the network to calculate the *earliest time* on which each event in the network can be achieved. The simplest way to determine earliest times is to add the activity duration to the previous event's early time. However, when two or more activities lead to the same event, you must apply the following rule.

RULE: When two or more activities enter a node, the earliest time when that event can be achieved is the *larger* of the durations on the paths entering the node. In all of the material that follows, we will restrict the meaning of the word *event* to mean the completion of all work that precedes it. We will therefore be determining the earliest and latest times that the preceding work can be completed without impacting the project end time.

NOTE: The time determined for the end or final event is the earliest finish for the project in working time (units can be hours, days, or weeks). Once weekends, holidays, and other breaks in the schedule are accounted for, the end *date* may be considerably later than the earliest finish determined by counting nothing but working times.

Conventional Analysis

By convention, in order to determine the latest time for each event, given the network logic and each activity duration, the final event late time is set equal to its earliest time, which has just been determined by the forward-pass computation. This is considered to be the best-case analysis, since it forces the project to end at its earliest possible time. If the project were allowed to run longer than the earliest possible completion, it would be stretching out the work unnecessarily. Of course, it may be that the project does not have to end until some later time, in which case that difference between the earliest possible finish and the required end date represent latitude that can be used to advantage. The project might be started later or fewer resources may be applied, allowing some activities to increase durations, but achieving lower labor costs (perhaps). In the example on page 115, the earliest time will be found to be 25 days, so the latest time will therefore be set equal to 25 days and then the backward-pass calculations will be made.

Backward-Pass Computations

A backward pass is made through the network to compute the *latest time* for each event in the network.

> **RULE:** When two or more activities leave a node, the latest time when that event can be achieved is the *smaller* of the durations of the paths leaving the node. (Remember, the event represents the completion of all work leading to it, so this will be the latest completion time for all preceding work.)

Note that the earliest finish for activity A is the earliest start for activities C and D. That is, $EF_A = ES_{C, D}$. Also the latest finish for A is the latest start for C. ($LF_A = LS_C$) However, it is not the latest start for D, as we shall see later.

The difference between LF_A and EF_A is called *slack*, which gives us a measure of latitude on the event. The physical meaning of slack is that the event can occur as early as day eight and as late as day 12 and the project can still be completed on time. Slack can be used in a number of ways, including to help manage resources better, as will be discussed later.

Note also activity D. The earliest that it can start is day eight, and the latest it can be finished is day 22. The distance between these events is 14 days, but the activity has a duration of only seven days. We say that activity D can *float* between those two event times for seven days (the difference between the distance and activity durations. This float is called *maximum float*. An equation for calculating maximum float follows.

$$\text{MAXIMUM FLOAT}_D = LF_D - ES_D - \text{DURATION}_D$$

Note also that if the earliest finish for event 4 is achieved and the latest finish for event 2 is allowed to occur, there is an eight-day distance between the two points. That means that activity D can float around for one day between those two points in time. This float is called *minimum float* or *free float*. This float is "free" to the activity: it will be available no matter what, so long as the event times specified are achieved. The equation for free float follows. (Not all writers use this same definition of free float. See Moder, Phillips, and Davis (1983) for other definitions.)

$$\text{MINIMUM FLOAT}_D = \text{EF}_4 - \text{LF}_2 - \text{DURATION}_D$$

That is, the minimum float is the Earliest Finish for the activity *j-event* minus the latest finish for the activity *i-event* minus the activity duration. Normally we will only be concerned with the maximum float available for an activity, and this is usually the number determined by software programs.

Next, it is instructive to examine activity D in more detail. Suppose activity D starts at its earliest time of day eight. How soon will it end? Since it has a duration of seven days, it can finish no earlier than day 15. The equation for this is:

$$\text{EF}_D = \text{ES}_D + D_{2,4}$$

Clearly the latest finish for activity D is day 22. But what is its latest start? If we take the latest finish for D and subtract the duration for D, we get the latest start, which is day 15. The equation is:

$$\text{LS}_D = \text{LF}_D - D_{2,4}$$

This shows that the latest time for event 2 is not the latest start for activity D, although it is for activity C.

Now examine path B-G. There is no float on either activity on this path. By convention, this is called the *Critical Path*, which means that if any of the work on this path falls behind schedule, then the end date will slip accordingly.

> When an activity has no float, it is called **critical**, since failure to complete work as scheduled will cause the end date to slip. Similarly, an event with no slack is called a **critical event**.

It is also informative to look at activity C. Suppose activity A slips to its latest time, which is day 12. How much float would then be available for activity C? It would be zero, since the late finish for C is day 15 and event 2 has occurred on day 12, making the time between events 2 and 3 only three days, and activity C has a duration of three days, so that it has no latitude. When this happens, a printout of the schedule would show activity C as now being critical, so that the network would have two critical paths. The remainder of activity B would still be critical, as would activity C and all of activity G.

Converting Arrow Diagrams to Bar Charts

While an arrow diagram is essential to do a proper analysis of the relationships between the activities in a project, and in order to find the longest path through the project, the best *working* tool is the bar chart. Those people doing the work will find it much easier to see when they are supposed to start and finish their jobs if you give them a bar chart. The arrow diagram above has been displayed as a bar chart (see Figure 9.5), making use of what was learned about the schedule from the network analysis. The trailing dots represent the activity float (they are not scaled one-dot-per-day).

Parkinson's Law: Work always expands to fit the time allowed.

As the chart shows, activities A & C have four days of float. Before any work is done on the project, this is true. However, as was just shown in the previous section, if activity A slips to its latest time of 12 days, then there is no float left for activity C. We say that the four days of float is the total *path float* on the path formed by activities A and C. It is shared between them. If none is used on A, then it is all available for C. However, if it is all used on A, then none is available for C.

This illustrates one of the pitfalls of drawing bar charts with float shown. If two different people are working on tasks A and C,

Figure 9.5. Bar Chart of Network in Figure 9.4.

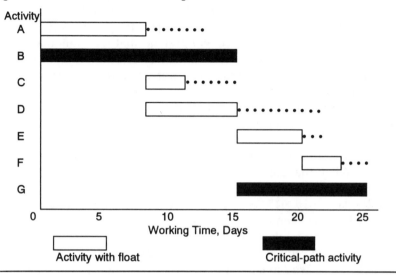

and they are unfamiliar with the network diagram, they may think that they each have four days of float, and that thinking can get them into trouble. It is considered by some project managers to be good practice to print bar charts with no float shown, and many software packages have this as an option.

> **Lewis' Law for Float: If you give them float, they'll take it.**

Constrained End Date

It is fairly common that an end date has been imposed on the project either by contract with the customer or by direction from management. This end date may be earlier than the earliest completion date determined by the forward-pass computation, in which case the schedule must be shortened somehow. This obviously means that the critical path must be shortened. However, other paths may be problems as well. If the network previously analyzed has the end constrained to a finish of 23 days, for example, we must determine what must be done to the network to achieve that result. (See Figure 9.6.) If we could start the project two days earlier, there would be no problem, but we will pretend that we cannot start earlier. Perhaps people or materials will not be available; or perhaps the specifications for the work will not be ready at that time. We must therefore compress the work.

With the end constrained to a 23-day completion, we must do a new backward-pass calculation to determine event late times. This is shown in Figure 9.7. We now have a situation in which the late times on events 1, 3, and 5 are *earlier than the early times* for the events. When the late time for an event is earlier than the early time for the event, we say that the slack is negative (for event 3, for example, slack = 13 − 15 = −2 days). When the slack becomes negative, we say that the event is *supercritical!*

By the same token, the float on activity B is now negative. Remember, maximum float is LF_B minus ES_B minus D_B. This is reasonable, since we have compressed the end date by two days, and since the critical path was the determinant of the earliest finish for the project, it must be two days too long, resulting in a "negative" latitude or float.

Figure 9.6. Constrained End Date.

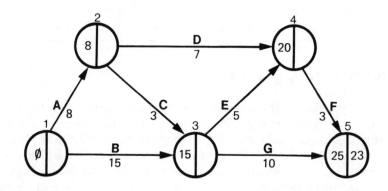

Figure 9.7. Project with Imposed End Date Earlier than Early Finish.

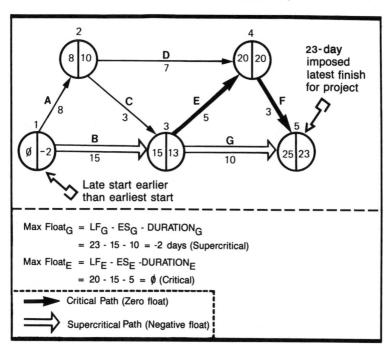

There is also a critical path at E-F, since the float there is now zero. Constraining the end date has created quite a mess of our network. As this analysis shows, we now know where we must make some adjustments if we are going to meet the constrained end date. For larger networks, this analysis is difficult to do manually, so it is very helpful to have the computer show where supercritical paths exist, so they can be fixed.

Solving the Problem

We know that the original critical path (B-G) must be shortened by two days. However, we also may need to do something with E-F. Where should we start?

One of the best operating rules of thumb is to begin making corrections as early in a network as possible. For that reason, we would try to shorten the duration of task B. The reason for this rule is that corrections made early are likely to propagate through the network. Further, if we take the time out early in the network, then as we actually do the job, if we find we need to shorten later tasks, we may still have some ability to do so, whereas if we have already trimmed them to the bare minimums, we won't have any latitude should the need arise.

How, then, can the duration of an activity be shortened? There are basically only four ways:

☞ *Add resources:* This can be done by adding people or working hours (overtime); of course, the costs will probably increase.

☞ *Reduce scope:* Don't do all of the work originally planned. This must be agreed to by the customer.

☞ *Change process:* This can be done by using a more efficient person or changing the working method. For example, spray painting is probably faster than roller painting.

☞ *Reduce quality:* While this is unsatisfactory (in my opinion) it is what sometimes is done. Project managers must guard against this happening.

The choice made will depend on a number of factors. For example, can resources be added? If the task is not labor-intensive, the answer may be negative. Writing software code or doing engineering design are two examples. Unless the work can be subdivided, adding resources will not make it go faster. On the contrary, it may go slower, because people will argue about how it should be done.

Painting walls can generally be made faster by the addition of resources, up to a point. That point occurs when you have so many painters that they get in each other's way. We call that the point of *diminishing returns.*

We might also redraw the network, somehow overlapping tasks B and G, but this is usually done only when other options have been explored and rejected.

Figure 9.8 shows the network with activity B shortened to 13 days duration. This does solve the problem, eliminating the super-critical path as well as getting E-F off the critical path. Whether it is feasible is another issue.

Figure 9.8. Activity B Shortened to Achieve 23-Day Completion.

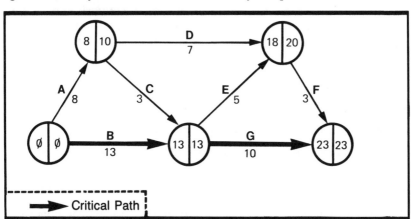

One caution is in order. Suppose we decide we will shorten the duration of activity B by having people work overtime. This may get us into trouble later on. The reason is that, when projects get into trouble—whether because of technical difficulty or other factors, it is commonplace to work more overtime in order to complete the job on time. What this means is that if we could meet the required time on activity B only by working overtime, and we now have to work overtime to solve a technical problem, we are almost certain to overload everyone to the point that we will miss the end date.

That leads to a rule of thumb that I suggest: try not to schedule a project so that significant overtime is required to meet original target dates, as this limits that option for resolving problems that occur later in the project.

I know that it is often necessary to violate this rule, but it is still a good one to apply when possible.

Handling Dual Critical Paths

When a network has two or more critical paths, the risk of slipping the end date increases. As a rule of thumb, a network should be manipulated until only one critical path remains. Following are some factors to consider in deciding which activities should have float.

Factors to consider in eliminating a dual critical path

Number of activities	Path with most activities.
Skill level of personnel	Path with least-skilled personnel probably most risky.
Technical risk	Path with greatest technical risk should have float.
Weather/uncontrollables	Give float to activities affected by uncontrollable factors.
Cost	Give float to activities that would have the greatest cost impact if they slip.
Historical data	No history—give float; Historically a problem—ditto.
Available backup plan	Give float to activities with no obvious backup.
Business cycle	If business tends to get hectic at times, try to give float to activities thereby affected.
Difficulty	Float for difficult activities.

Chapter 10

Scheduling with Resource Constraints

The Assumption of Unlimited Resources

All of the scheduling computations in the previous chapter were made assuming that the activity durations were achievable. However, the time required to complete an activity depends on the resource(s) assigned to it, and if the level of planned resources is not available, then the work cannot be completed as planned. Assumption three of project scheduling was that we have unlimited resources. Naturally that is not the case, even for the largest organizations.

Elsewhere I have said that project scheduling cannot be successful unless the project manager can solve the resource allocation problem. Every organization has a fixed number of resources, which are shared among all projects. Also, while basic schedule computations can be made manually—even for fairly sizable networks, the resource allocation problem quickly grows to such proportions that it can only be solved with a computer. Therefore, this chapter will illustrate how a schedule is developed to account for the availability of resources, using Project Workbench^{TM}, produced by Applied Business Technologies. Other PC software may handle the problem a bit differently, but the approach illustrated here will give a general outline that will be representative of most of the software available today.

The Effect of Limited Resources on Schedule Float

When resource limitations exist, the float that was determined through conventional critical-path analysis may have to be used to avoid overloading resources. To illustrate the resource allocation approach, the network developed in the previous chapter will be used, with resources assigned to each activity.

When a number of people can all do the same work, they are treated as a "pool" of resources. The initial analysis presented here will make use of pooled resources. That analysis will be followed by one in which specific individuals are assigned to the tasks.

In Chapter 7, on estimating, we said that we begin estimating an activity duration by assuming that a certain kind of resource will be applied. For example, a ten-year-old boy probably cannot mow grass with a push mower as fast as a sixteen-year-old. So if I am going to estimate how long it will take to cut the grass, I must begin by deciding who will do it. Then I know how long it will take.

When this approach is used, there is another assumption involved. It is that the task has a fixed duration. This is the most usual way in which activity durations are handled, so that system will be illustrated first. However, if a task duration depends on how many resources are applied, we may be able to achieve a better schedule if we move resources around, changing activity durations. This is the *variable-duration* approach, and that approach will be illustrated as well. The network used for computations is repeated in Figure 10.1 for convenience.

Figure 10.1. Network to Illustrate CPM Analysis.

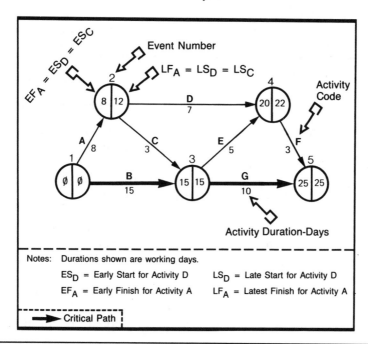

Scheduling with Fixed-Duration Activities

To allocate resources, the level of resources *available* to the project must be specified, together with the level *required* by each activity. The software then attempts to schedule work so that the available resources are not overloaded.

The resource level available is measured by taking the number of people, multiplying by the total amount of time they are going to work, and specifying the product as the availability. If two individuals are available for 40 hours per week, we say that 80 person-hours per week are available. If time is measured in days, then we say that for a five-day week, with two people, we have 10 person-days per week available. These are the units that will be used in the following analysis.

For the project analyzed here, two categories of workers were specified, in the levels shown below:

Description of Resource	Number Available	Person-Days/Week
Senior Worker	2	10
Helper	1	5

Two senior workers are available. As the table shows, if a normal five-day work week is assumed, then there are 10 person-days of labor available to apply to the project each week. For the Helpers, one is available, yielding five person-days of labor each week.

Note that holidays, vacations, and overtime will affect the amount of labor available during the period in which they occur, so that the schedule will reflect a total *elapsed* time different than that obtained if a constant level of labor were available.

For the analysis that follows, no vacations have been entered, but standard legal holidays are left in the program to show the effect of having days on which no one works. (Workbench has a list of standard U.S. holidays already entered into the software. The list can be modified to meet the needs of individual organizations or countries.)

As each activity is entered into the computer, a level of resources is assigned to each task. The levels assigned are as shown in the table.

Activity	Duration	Senior Worker	Helper
A	8	8	0
B	15	15	3
C	3	0	3
D	7	7	2
E	5	5	0
F	3	3	1
G	10	20	10

Note: Durations are in days; allocations in person-days total.

Scheduling with Resource Constraints

In the previous chapter, we saw that this network required 25 working days for completion, assuming that the required number of resources were available. The first Project Workbench analysis is made in that manner, and shows a 25-day working schedule, as was determined in Chapter 9. This schedule is shown in Figure 10.2. Using June 17, 1991, as a starting date, the project will end on July 22. As the schedule shows, because July 4 is a holiday, it is dropped from working time, so that the actual end date is 26 days from project start. A network diagram produced by Project Workbench is also shown in Figure 10.3.

Figure 10.2. Project Workbench Schedule, Unleveled.

```
──────────────── Project description ────────────────
Report: Gantt Chart          Date:  6-17-91         Time: 14:32

Title: Project Zero          Project ID: SP         Version: 1

Manager: Jim Lewis           Project filename: ZERO6HBK Dept:

Project start: 6-17-91       Project end:           Budget:

Description:  Small Network to illustrate resource scheduling
──────────────────────── Legend ────────────────────────
 █████  Activity                    ♦       Milestone
 C████  Activity on critical path   L████   Locked activity
 ═███   Partially completed activity ═══    Completed activity
 ▲      Original start date          ▼      Original end date
 .....  Discontinuous activity      .....♦  Baselined milestone
```

						July 1991				August 1991	
Project Zero	Day	Resrc	17	24	1	8	15	22	29	5	12	19
Activity A	8	SW	████████									
Activity B	18	SW HE	C███████████ ██									
Activity C	3	HE			███							
Activity D	9	SW HE			██████ ██							
Activity E	5	SW				██████						
Activity F	4	SW HE				·████						
Activity G	30	SW HE				·C███████						
Finish	♦	X					C♦					

						July 1991				August 1991	
Project Zero	Day	Resrc	17	24	1	8	15	22	29	5	12	19

Resource summary
Utilization

UNASSIGNED		X							
Senior Work	10.0	SW	10.0	10.0	8.0	14.0	14.0	2.0	
Helper	5.0	HE	1.0	3.6	3.0	4.4	6.0	1.0	
Total days			11.0	13.6	11.0	18.4	20.0	3.0	

Figure 10.3. Network Diagram from Project Workbench.

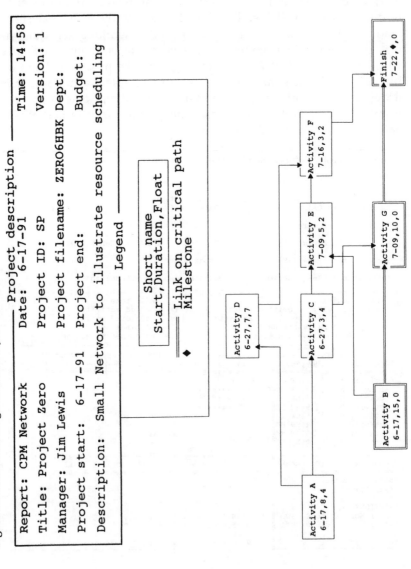

```
 ┌─────────────────── Project description ───────────────────┐
 │ Report: CPM Network        Date: 6-17-91        Time: 14:58 │
 │ Title: Project Zero        Project ID: SP       Version: 1  │
 │ Manager: Jim Lewis         Project filename: ZERO6HBK Dept: │
 │ Project start: 6-17-91     Project end:         Budget:     │
 │ Description: Small Network to illustrate resource scheduling│
 └──────────────────────────── Legend ───────────────────────┘

              ┌─────────────────────────────┐
              │        Short name           │
              │   Start,Duration,Float      │
              │   ─── Link on critical path │
              │   ◆ Milestone               │
              └─────────────────────────────┘
```

Further, underneath the schedule is a listing of the level of resources required during each week. In Project Workbench terminology, this is called a Resource Spreadsheet™· Since we have 10 person-days per week of senior workers available and five person-days of helper, if the actual utilization does not exceed those levels, then the project is workable.

However, there are two weeks in which the senior worker level goes to 14 and one week in which the helper level reaches six, so the resources are overloaded. This tells us the level of resources that would be needed in order to achieve the earliest possible completion for the project, if the network is left as originally drawn.

On the other hand, we may not have enough resources to meet the required levels, so we might ask when the project will end if we maintain the availability as originally specified and do not overload anyone. (Working overtime is a way of increasing availability of resources, so we will momentarily consider that option unsuitable and do another analysis.)

When the first analysis is conducted, we are essentially considering time to be critical, so all activities are scheduled to start at their earliest times, and this will cause the project to end at its earliest possible time. If we want to avoid overloading resources, we call the analysis a *resource-critical allocation*. Under this analysis, the end date may slip beyond July 22, in order to avoid overloading people.

This is indeed what happens, as shown in Figure 10.4. The end date slips out to August 6. To understand exactly what is happening requires some study. First of all, the resource spreadsheet levels portrayed beneath the Gantt chart are weekly totals and do not show enough detail at times. For that reason, I have manually plotted resource utilization levels on a daily basis for the schedule in its initial form—that is, with the July 22 end date being met. These histograms are shown in Figures 10.5 and 10.6.

The histogram for senior workers shows that the senior workers are only overloaded from July 9 through July 18, because activity G requires two senior workers alone, while E and F each require one senior worker, making the total requirement three if the activities are scheduled to start at their earliest times. Project Workbench gives first priority to the critical-path activity G and delays E and F until G has finished, thereby slipping the end date. Although Work-

Figure 10.4. Bar Chart after Auto Scheduling.

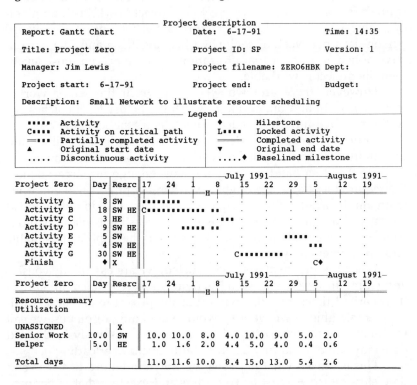

bench does not show that E and F are critical tasks, we know that once an activity has used up all of its float, it is indeed critical. This means that the critical path under resource limitations is not the same as under ideal conditions, a fact that has gotten countless projects in trouble.

As for the senior worker, we can see that there are overloads on June 27 through 29 and again on July 16 through 18. The overload beginning on June 27 causes task C to slip until critical-path task B has been completed. This is again because the critical path is always given first priority for resources. If this is not done, we

Figure 10.5. Loading the Senior Worker.

Figure 10.6. Loading of the Helper.

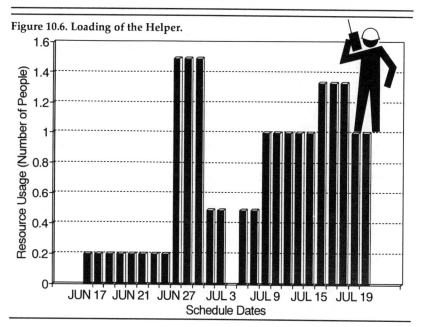

know for certain that the project end date is certain to slip, since anything that delays a critical task delays the end date accordingly.

The reason for the problem has to do with how the helper is assigned to the tasks. Activity B is scheduled to take 15 working days, with the senior worker clearly carrying most of the load. The helper is allocated only three days on the task, and the question is how that allocation is made.

Naturally, there are many ways such a fractional allocation can be made. Some of the more common ones are:

☞ Three full days at the front or back of the activity. An example would be that the person comes in and does some setup work for the senior worker, or perhaps runs tests at the end of the activity.

☞ One full day each week. Perhaps the helper does preparation or cleanup work for the senior worker.

☞ Uniform fractional loading of 20 percent per day for the 15-day duration. This is fairly common for a helper. It amounts to about 1.6 hours per day.

For simplicity, I assigned the helper to task B at a uniform level of 20 percent per day. This causes the helper to be overloaded when task C begins, since this task is done entirely by the helper. And, since the critical path is given priority, task C is simply delayed until task B is completed, resulting in a three-day impact on the project end date.

In a "real-world" project, there are a couple of ways this might be handled. One would be for the senior worker to give up the helper for the three days it takes to do activity C. Another would be to work the helper 1.6 hours per day of overtime, which would not be too bad. Either of these is a potential, and it is completely within the capability of Workbench to do either. Activity B can have the helper variably loaded, so that no helper is used during the time when C is to be done, thus pulling the project end date in by three days.

Another solution is to front-load activity B with the helper. That is, use the helper for three full days at the beginning of B, so that by the time C starts, the helper is no longer needed on B. This solution is shown in Figure 10.7.

Now the only cause of the end date being impacted is the senior workers being overloaded with E and F parallel with G, which means that extra workers will be required or the end date is going to slip. And, while we could increase availability by working overtime, it will require a 40 percent increase, which means 7-day weeks for two weeks to get the job done.

Figure 10.7. Helper Front-Loaded on Activity B.

```
─────────────── Project description ───────────────
Report: Gantt Chart          Date:  6-17-91              Time: 14:40

Title: Project Zero          Project ID: SP              Version: 1

Manager: Jim Lewis           Project filename: ZERO6HB1 Dept:

Project start:  6-17-91      Project end:               Budget:

Description:  Small Network to illustrate resource scheduling
─────────────────────────── Legend ───────────────────────────
 ▪▪▪▪▪  Activity                  ◆       Milestone
 C▪▪▪▪  Activity on critical path  L▪▪▪▪   Locked activity
 ═▪▪▪   Partially completed activity ═══   Completed activity
 ▲      Original start date        ▼       Original end date
 .....  Discontinuous activity     .....◆  Baselined milestone
```

			─July 1991─					─August 1991─	
Project Zero	Day	Resrc	17 24	1 8	15 22	29	5	12	19
Activity A	8	SW	▪▪▪▪▪▪▪▪						
Activity B	18	SW HE	C▪▪▪▪▪▪▪▪▪▪▪ ▪▪						
Activity C	3	HE	. ▪▪▪						
Activity D	9	SW HE	▪▪▪▪▪▪ ▪▪						
Activity E	5	SW			. ▪▪▪▪▪▪				
Activity F	4	SW HE				▪▪▪			
Activity G	30	SW HE			·C▪▪▪▪▪▪▪▪▪				
Finish	◆	X				C◆			

			─July 1991─					─August 1991─	
Project Zero	Day	Resrc	17 24	1 8	15 22	29	5	12	19

```
Resource summary
Utilization

UNASSIGNED      |     | X
Senior Work     |10.0 | SW    10.0 10.0  8.0 10.0 10.0  6.0  4.0
Helper          | 5.0 | HE     3.0  2.0  2.5  4.5  5.0  1.0  1.0

Total days      |     |       13.0 12.0 10.5 14.5 15.0  7.0  5.0
```

Scheduling with Variable-Duration Activities

In the resource analyses made so far, we have assumed that the activity would take a fixed amount of time with a certain level of resources applied. Under these conditions, the scheduling of activities becomes an "all-or-nothing" situation. If a person were available half-time to work on a second task, but the second task was specified as requiring a full-time person, then the computer would delay the second task until the person became available full-time. This is, of course, not an effective allocation, and in practice a project manager would start on the second task on a part-time basis.

Since the duration of an activity is normally a function of the amount of labor applied, then it might be possible to achieve a shorter schedule by allowing the software to apply whatever level of resources are available, and adjust the activity durations to "fit." This is equivalent to treating the activity duration as *variable*. Project Workbench has such a feature.

To illustrate, the activities in the network were changed to variable-duration types, but resource availability and resources required for each activity were left unchanged. The helper is front-loaded on activity B, which we saw in the previous analysis prevents activity C from slipping the end date by three days. The new analysis is shown in Figure 10.8.

As the bar chart shows, the schedule is pulled in from an August 1 end date to July 29, an improvement of three days. This is still not as good as the ideal case, which shows a July 22 finish, but it is an improvement in both the end date and utilization of resources.

All of the activity durations have been changed by the computer. Table 10.1 shows a comparison of original and new durations.

Table 10.1. Comparison of Original and New Durations.

Activity	A	B	C	D	E	F	G
Original Duration	8	15	3	7	5	3	10
New Duration	4	13	4	15	3	2	11

Figure 10.8. Variable Duration Activities.

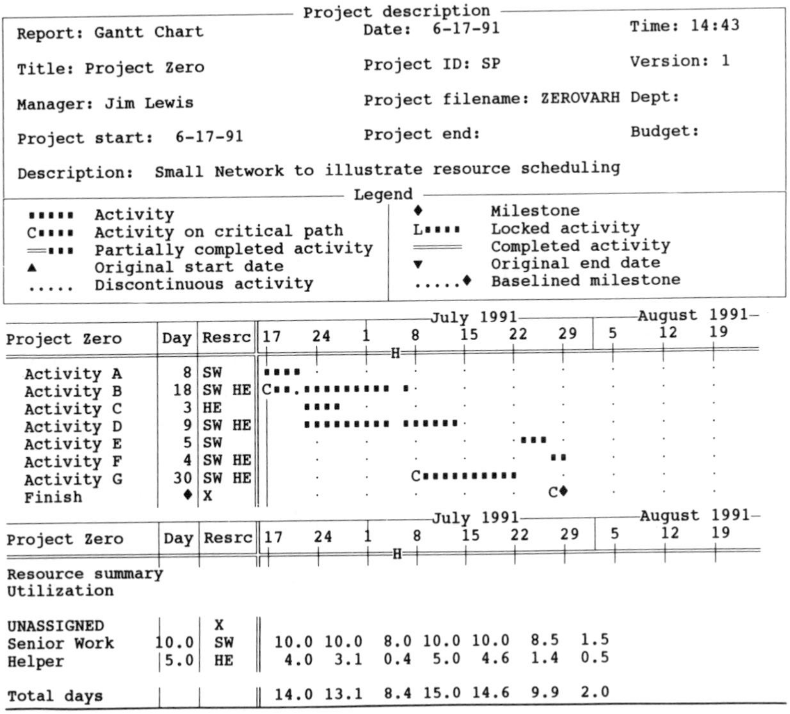

```
──────────────────── Project description ────────────────────
Report: Gantt Chart          Date:  6-17-91          Time: 14:43

Title: Project Zero          Project ID: SP          Version: 1

Manager: Jim Lewis           Project filename: ZEROVARH Dept:

Project start:  6-17-91      Project end:            Budget:

Description:  Small Network to illustrate resource scheduling
──────────────────────────── Legend ────────────────────────────
 ▪▪▪▪▪  Activity                     ◆       Milestone
 C▪▪▪▪  Activity on critical path    L▪▪▪▪   Locked activity
 ═▪▪▪   Partially completed activity  ═══    Completed activity
 ▲      Original start date          ▼       Original end date
 .....  Discontinuous activity       .....◆  Baselined milestone
```

			─────────────July 1991─────────────── ──August 1991──
Project Zero	Day	Resrc	17 24 1 8 15 22 29 5 12 19
			H
Activity A	8	SW	▪▪▪▪▪ .
Activity B	18	SW HE	C▪▪.▪▪▪▪▪▪▪▪▪ ▪.
Activity C	3	HE	▪▪▪▪▪
Activity D	9	SW HE	▪▪▪▪▪▪▪▪▪ ▪▪▪▪▪▪.
Activity E	5	SW	▪▪▪▪
Activity F	4	SW HE	▪▪
Activity G	30	SW HE	C▪▪▪▪▪▪▪▪▪▪▪
Finish	◆	X	C◆

			─────────────July 1991─────────────── ──August 1991──
Project Zero	Day	Resrc	17 24 1 8 15 22 29 5 12 19
			H

Resource summary
Utilization

		X	17 24 1 8 15 22 29
UNASSIGNED		X	
Senior Work	10.0	SW	10.0 10.0 8.0 10.0 10.0 8.5 1.5
Helper	5.0	HE	4.0 3.1 0.4 5.0 4.6 1.4 0.5
Total days			14.0 13.1 8.4 15.0 14.6 9.9 2.0

To understand what the computer has done, it is necessary to study the resource loading on each activity, which unfortunately can only be done on the screen, since the histograms cannot be printed. Activity A, which originally had a duration of eight days, now has a duration of four days. This can only be possible if the level of senior workers is doubled, which is what the computer has done. Further, the computer has "assumed" that activity C can start with the helper only, working the first three days. So activity C has a one-day interruption, while activity A is being completed, whereupon it resumes, but the level of senior workers applied is actually

1.5 per day, which allows the activity to be completed in the remaining 12 days. At the same time, activity D has now been stretched out to a 15-day duration because the computer has allocated the remaining 0.5 senior worker available to that activity and (a little more than) doubled its duration. Whether this would be an acceptable "real-world" solution would have to be decided by the project manager. The computer has simply given a *potential* solution.

Scheduling with Specific Individuals

When resources cannot be considered "generic" or "pooled," the resulting schedule may be different than that obtained when pooled resources are used. Using the same network as before, let us assume that the two senior-level people available to work on the project are named Tom Trump and Sue Simms. In addition, they have a helper named Charlie Clark. For the analysis that follows, we will also assume that the senior workers have loaded labor rates of $240 per day and the helper's rate is $160 per day. (Loaded labor rate is direct salary plus overhead. This is the figure used by most managers to calculate total project labor costs.) By entering labor rates into the resource tables, Project Workbench™ will calculate total labor costs for the project.

The assignments are made as in the previous analyses, with Tom Trump and Sue Simms being applied for the senior workers and Charlie Clark for the helper. In Figure 10.9 is a standard CPM analysis in which the project is scheduled to end at its earliest possible time of July 22. The listings beside each activity show who is assigned to each activity, and the resource levels under the schedule show the loading that would be necessary to complete the project by July 22. Since a level greater than 5.0 represents more than five working days required in a week, it is easy to see that Sue Simms is overloaded during the week of July 8 and all three are overloaded during the next week.

Next an analysis was conducted using Autoschedule to level resource usage, and the result is shown in Figure 10.10. The end date is August 1, as was determined using pooled resources, the

Figure 10.9. Schedule with Individuals Assigned.

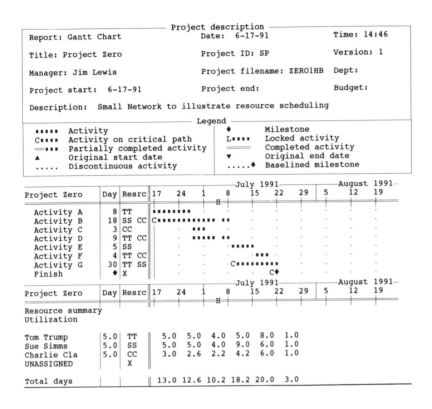

```
─────────────────── Project description ───────────────────
Report: Gantt Chart              Date:  6-17-91            Time: 14:46

Title: Project Zero             Project ID: SP            Version: 1

Manager: Jim Lewis          Project filename: ZERO1HB  Dept:

Project start:  6-17-91         Project end:             Budget:

Description:  Small Network to illustrate resource scheduling
────────────────────────────── Legend ──────────────────────────
 ■■■■■  Activity                    │   ◆      Milestone
 C■■■■  Activity on critical path   │   L■■■■  Locked activity
 ═■■■   Partially completed activity│   ══════ Completed activity
 ▲      Original start date         │   ▼      Original end date
 .....  Discontinuous activity      │   .....◆ Baselined milestone
──────────────────────────────────────────────────────────────────
                                    ──July 1991──      ──August 1991──
Project Zero     Day Resrc 17   24   1    8   15   22   29   5   12   19

Activity A        8  TT    ■■■■■■■■ .    .    .    .    .    .    .    .
Activity B       18  SS CC C■■■■■■■■■■■■ ■■   .    .    .    .    .    .
Activity C        3  CC         .   ■■■  .    .    .    .    .    .    .
Activity D        9  TT CC      .   ■■■■■■ ■■ .    .    .    .    .    .
Activity E        5  SS         .    .  ..■■■■■■   .    .    .    .    .
Activity F        4  TT CC      .    .    . ■■■ .  .    .    .    .    .
Activity G       30  TT SS      .    .    .·C■■■■■■■■■■■  .    .    .    .
Finish            ◆  X          .    .    .    .   C◆    .    .    .    .
                                    ──July 1991──      ──August 1991──
Project Zero     Day Resrc 17   24   1    8   15   22   29   5   12   19

Resource summary
Utilization

Tom Trump      5.0  TT    5.0  5.0  4.0  5.0  8.0  1.0
Sue Simms      5.0  SS    5.0  5.0  4.0  9.0  6.0  1.0
Charlie Cla    5.0  CC    3.0  2.6  2.2  4.2  6.0  1.0
UNASSIGNED          X

Total days                13.0 12.6 10.2 18.2 20.0  3.0
```

reason being simply that the individuals have been assigned to the project in the same way as was done using pooled people. In Figure 10.11 is an activity listing that shows start and finish dates for each task, the maximum float available under initial conditions, and the total labor costs for each activity. According to this printout, the total labor costs for the project will be $16,960.

It is clear that the reason that the end date is slipping is that the senior workers are overloaded on tasks E, F, and G. If we could

Figure 10.10. Schedule wtih Resources Leveled.

```
──────────────────────── Project description ────────────────────────
 Report: Gantt Chart              Date:  6-17-91              Time: 14:48

 Title: Project Zero              Project ID: SP              Version: 1

 Manager: Jim Lewis               Project filename: ZERO1HB   Dept:

 Project start:  6-17-91          Project end:                Budget:

 Description:  Small Network to illustrate resource scheduling
────────────────────────────── Legend ──────────────────────────────
 ▪▪▪▪▪  Activity                      ◆         Milestone
 C▪▪▪▪  Activity on critical path     L▪▪▪▪     Locked activity
 ══▪▪▪  Partially completed activity  ═══       Completed activity
   ▲    Original start date           ▼         Original end date
 .....  Discontinuous activity        .....◆    Baselined milestone
```

Project Zero	Day	Resrc	┃17 24 1 8 15 22 29 5 12 19
			┃ ─────July 1991──── ──August 1991─
Activity A	8	TT	▪▪▪▪▪▪▪▪
Activity B	18	SS CC	C▪▪▪▪▪▪▪▪▪▪▪▪ ▪▪
Activity C	3	CC	▪▪▪
Activity D	9	TT CC	▪▪▪▪▪ ▪▪
Activity E	5	SS	▪▪▪▪▪
Activity F	4	TT CC	▪▪▪
Activity G	30	TT SS	C▪▪▪▪▪▪▪▪▪
Finish	◆	X	C◆

Project Zero	Day	Resrc	┃17 24 1 8 15 22 29 5 12 19
			┃ ─────July 1991──── ──August 1991─
Resource summary Utilization			

			17	24	1	8	15	22	29	5	12	19
Tom Trump	5.0	TT	5.0	5.0	4.0	5.0	5.0	1.0	3.0			
Sue Simms	5.0	SS	5.0	5.0	4.0	5.0	5.0	5.0	1.0			
Charlie Cla	5.0	CC	3.0	2.0	2.5	4.5	5.0	1.0	1.0			
UNASSIGNED		X										
Total days			13.0	12.0	10.5	14.5	15.0	7.0	5.0			

assign another person to do activities E and F, we might be able to meet the July 22 end date.

To see if this will work, Mary Martin was assigned to the project and allocated only to Activities E and F. When Autoschedule was run using this configuration, only Charlie Clark was overloaded during the week of July 15. If he is removed from Activity F, assuming that Mary Martin can do that task by herself, then an

Figure 10.11. Activity Listing with Labor Costs.

```
                          ┌──────────── Project description ────────────┐
  Report: Activity Detail │            Date:  6-17-91          Time: 14:51 │
  Title: Project Zero     │      Project ID: SP              Version: 1    │
  Manager: Jim Lewis      │      Project filename: ZERO1HB   Dept:         │
  Project start:  6-17-91 │      Project end:               Budget:        │
  Description:  Small Network to illustrate resource scheduling            │
  └───────────────────────────── Revised Dates and Usage ─────────────────┘
```

----------------Name----------------	Start	End	Duration Bus days	Status	--------Resource assignments-------- Name	Usage	Cost	Float	Pty	Category
Activity A	6-17-91	6-26-91	8		Tom Trump	8.0	1,920	4		
Activity B	6-17-91	7-08-91	15		Sue Simms	15.0	3,600	0		
					Charlie Clark	3.0	480			
Activity C	6-27-91	7-01-91	3		Charlie Clark	3.0	480	4		
Activity D	6-27-91	7-08-91	7		Tom Trump	7.0	1,680	7		
					Charlie Clark	2.0	320			
Activity E	7-23-91	7-29-91	5		Sue Simms	5.0	1,200	2		
Activity F	7-30-91	8-01-91	3		Tom Trump	3.0	720	2		
					Charlie Clark	1.0	160			
Activity G	7-09-91	7-22-91	10		Tom Trump	10.0	2,400	0		
					Sue Simms	10.0	2,400			
					Charlie Clark	10.0	1,600			
Finish		8-01-91			UNASSIGNED			0		
Total project	6-17-91	8-01-91	33		Total days	77.0				
					Total cost		16,960			

----------------Name---------------- Resource summary		Units	Amount	Start	End	Duration Bus days	Status	--------Resource assignments-------- Name	Usage	Cost	Float	Pty	Category
Name													
Tom Trump		days	240	6-17-91	8-01-91			TT	28.0	6,720			
Sue Simms		days	240	6-17-91	7-29-91			SS	30.0	7,200			
Charlie Clark		days	160	6-17-91	8-01-91			CC	19.0	3,040			
UNASSIGNED		days		8-01-91	8-01-91			X		0			
Total project				6-17-91	8-01-91	33		Total days	77.0				
								Total cost		16,960			

Autoschedule shows that no one would be overloaded, holding the July 22 end date. This schedule is shown in Figure 10.12. Finally, in Figure 10.13 is an activity listing, which shows that the total project costs have dropped to $16,480, because Charlie Clark has been removed from activity F and Mary Martin is doing activities E and F at a lower labor rate than was originally planned with Tom Trump and Sue Simms doing them.

This analysis shows what is possible through the use of contemporary scheduling software. In planning a project, the computer

Figure 10.12. Scheduled with Mary Martin Added.

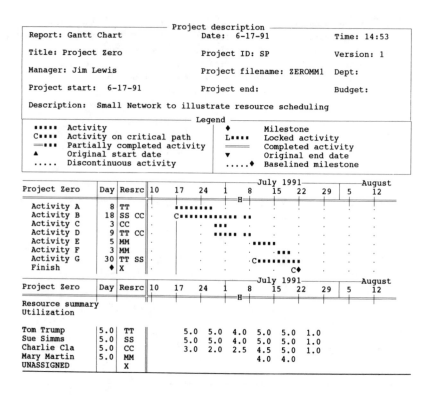

```
------------------------- Project description -------------------------
  Report: Gantt Chart              Date:  6-17-91              Time: 14:53

  Title: Project Zero              Project ID: SP              Version: 1

  Manager: Jim Lewis               Project filename: ZEROMM1   Dept:

  Project start:  6-17-91          Project end:                Budget:

  Description:  Small Network to illustrate resource scheduling
-------------------------------- Legend --------------------------------
  ■■■■■  Activity                        ♦         Milestone
  C■■■■  Activity on critical path       L■■■■     Locked activity
  ═■■■   Partially completed activity    ═════     Completed activity
  ▲      Original start date             ▼         Original end date
  .....  Discontinuous activity          .....♦    Baselined milestone
```

Project Zero	Day	Resrc	10	17	24	1	8	15	22	29	5	12

July 1991 ———— August

Activity A	8	TT
Activity B	18	SS CC
Activity C	3	CC
Activity D	9	TT CC
Activity E	5	MM
Activity F	3	MM
Activity G	30	TT SS
Finish	♦	X

Project Zero	Day	Resrc	10	17	24	1	8	15	22	29	5	12

July 1991 ———— August

Resource summary
Utilization

Tom Trump	5.0	TT	5.0	5.0	4.0	5.0	5.0	1.0		
Sue Simms	5.0	SS	5.0	5.0	4.0	5.0	5.0	1.0		
Charlie Cla	5.0	CC	3.0	2.0	2.5	4.5	5.0	1.0		
Mary Martin	5.0	MM				4.0	4.0			
UNASSIGNED		X								

allows various *what-if* analyses to be performed quickly, so that an acceptable solution can be devised. Once this has been done, the schedule becomes the plan for the project and work can then be tracked against it. How this is done will be covered in the chapters on control that follow.

Figure 10.13. New Project Cost Figures.


```
                              Project description
  Report: Activity Detail              Date:  6-17-91              Time: 14:56
  Title: Project Zero            Project ID: SP                Version: 1
  Manager: Jim Lewis             Project filename: ZEROMM1  Dept:
  Project start:  6-17-91           Project end:              Budget:
  Description:  Small Network to illustrate resource scheduling
```

Revised Dates and Usage

------Name------	Start	End	Duration Bus days	Status	Resource assignments Name	Usage	Cost	Float Pty Category
Activity A	6-17-91	6-26-91	8		Tom Trump	8.0	1,920	4
Activity B	6-17-91	7-08-91	15		Sue Simms	15.0	3,600	0
					Charlie Clark	3.0	480	
Activity C	6-27-91	7-01-91	3		Charlie Clark	3.0	480	4
Activity D	6-27-91	7-08-91	7		Tom Trump	7.0	1,680	7
					Charlie Clark	2.0	320	
Activity E	7-09-91	7-15-91	5		Mary Martin	5.0	1,000	2
Activity F	7-16-91	7-18-91	3		Mary Martin	3.0	600	2
Activity G	7-09-91	7-22-91	10		Tom Trump	10.0	2,400	0
					Sue Simms	10.0	2,400	
					Charlie Clark	10.0	1,600	
Finish		7-22-91			UNASSIGNED			0
Total project	6-17-91	7-22-91	25		Total days	76.0		
					Total cost		16,480	

------Name------			Start	End	Duration Bus days	Status	Resource assignments Name	Usage	Cost	Float Pty Category
Resource summary										
Name	Units	Amount								
Tom Trump	days	240	6-17-91	7-22-91			TT	25.0	6,000	
Sue Simms	days	240	6-17-91	7-22-91			SS	25.0	6,000	
Charlie Clark	days	160	6-17-91	7-22-91			CC	18.0	2,880	
Mary Martin	days	200	7-09-91	7-18-91			MM	8.0	1,600	
UNASSIGNED	days		7-22-91	7-22-91			X		0	
Total project			6-17-91	7-22-91	25		Total days	76.0		
							Total cost		16,480	

Multiproject Scheduling

The impact of resource constraints illustrated by the single-project example above is magnified in scheduling multiple projects—situations where several separate, independent projects are linked together through their dependence upon a pool of common resources.

Project Workbench™ and other software packages allow the multiproject situation to be analyzed in much the same way as was done for a single project.

Practical Suggestions for Resource Allocation

☞ Don't allocate resources to be available on the project more than 80 percent of the time. Reduce availability even more if the person has a number of other projects to support.

☞ Read the software manual carefully to learn what kind of allocation rule(s) are being applied.

☞ Begin by assigning a *generic* resource to the project and then allocate a specific individual.

Chapter 11

Scheduling with PERT*

PERT Compared with CPM

When a project consists of activities, most of which are similar to others that have been performed a large number of times, CPM scheduling is generally used. With CPM, estimates of activity durations are based on historical data, and are assumed to be the mean or average time that the activity has taken to perform in the past.

---

*   This chapter is adapted from Chapter 8 of my book *Project Planning, Scheduling, and Control,* Chicago,Probus, 1991

However, when a project contains a majority of activities for which no experience exists—that is, no historical data is available—then the estimating difficulty becomes significant. When no experience is available to use as a guide, the only thing that can be done is make the best possible guess, based on *whatever* relevant experience one has.

It seems clear, however, that the more unique an activity is, the less certain the estimate of its duration, and therefore, the more *risky* the project will be in terms of control. And since a lot of projects (such as research and development) fall into this category, the question naturally arises of whether there might not be some method that could be employed to reduce estimating risk.

It was in response to this problem that PERT was developed around 1958 as a joint effort between the Unites States Navy and the Booz, Allen, and Hamilton consulting firm, and originally applied to the Polaris submarine project.

While estimates of activity durations for CPM projects are taken as averages based on history, once they are in place, they are often assumed to be more-or-less fixed, or to use the colloquial expression, they are "engraved in granite."

The PERT system, however, is based on the recognition that estimates are uncertain, and therefore it makes sense to talk of *ranges* of durations, and the *probability* that an activity duration will fall into that range, rather than assuming that an activity will be completed in a fixed amount of time.

Empirical Frequency Distributions

To understand the probability and statistics involved in PERT, consider an activity that has been performed in the past many times under essentially the same conditions. For the activity in question, duration times ranged from seven to 17 days. Now suppose that you count the number of times the activity required seven days to perform, eight days to perform, etc., and you display the resulting information in the form of an empirical frequency distribution, or histogram, as shown in Figure 11.1.

Figure 11.1. Empirical Frequency Distribution.

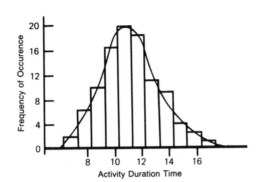

As we know from statistics, if an infinite number of observations were made, the width of the intervals in this figure approaches zero, and the distribution would merge into some smooth curve. This type of curve is the theoretical probability density of the random variable. The total area under such a curve is made to be exactly one, so that the area under the curve between any two values of **t** is directly the probability that the random variable **r** will fall in this interval. When this is done, the curve is called a *normal distribution curve*. It is also often called a bell-shaped curve.

Once the normal distribution curve exists for an activity, it is then a simple matter to extract the average-expected activity duration from the curve and to use that time as the estimate for how long the work will take.

However, under those conditions when no such distribution exists, we could still say that the problem is to arrive at our *best approximation of what the average expected duration would be if we could* perform the work over and over to develop the normal distribution curve. It is the answer to this question that forms the heart of the PERT system.

PERT System of Three Time Estimates

Even though a project may consist of activities for which little or no experience exists, most planners will have some relevant experience, so in most cases it is possible to make an educated "guess" of the *most likely* time the work will take. In addition, estimates can be made of how long the work would take if things go better than expected and, conversely, if things go worse than expected. These are called the *optimistic* and *pessimistic* conditions, respectively.

They are not defined as best-and worst-case, however. See the definitions that follow for the exact meanings of the terms *optimistic* and *pessimistic* in Table 11.1.

Table 11.1. Terms Used in PERT Estimating.

Definitions:

a = **optimistic performance time:** the time that would be improved only one time in twenty, if the activity could be completed repeatedly under the same essential conditions.

m = **most likely time:** the modal value of the distribution, or value which is most likely to occur more often than any other value.

b = **pessimistic performance time:** the time that would be exceeded only one time in twenty if the activity could be performed repeatedly under the same essential conditions.

These three estimates can be thought of as representing aspects of the normal distribution curve that could be developed if the work were performed a sufficient number of times. Another way to think of them is to say they represent *information* or data about the work in question. Taken together, perhaps a computation of the distribution *mean* can be made.

This is the essence of the PERT system, although what has been presented is an admittedly simplified presentation. The interested reader should consult Moder, Phillips, and Davis (1983) for a more thorough treatment of the statistics involved. For our purposes, all that matters is the application of the method.

PERT Computations

In order to combine the three estimates to calculate the expected mean duration for the activity, a formula was derived, based on principles from statistics. The estimate of average expected time to perform an activity is given by the following expression:

$$t_e = \frac{a + 4m + b}{6}$$

where:

t_e = expected time
a = optimistic time estimate
m = most likely time
b = pessimistic time

These values of t_e are used as the durations of activities in a PERT network. Given those estimated durations, the network calculations are identical to those for CPM. A forward-pass computation yields earliest times for events and a backward-pass provides latest times.

Estimating Probability of Scheduled Completion

What is gained by PERT, compared to CPM, is the ability to now compute a *confidence interval* for each activity and for the critical path, once it has been located. To do this, the standard deviation of each activity distribution must be known. With PERT software, such a computation would be automatically made by the software. However, if CPM software is used to do scheduling, the calculations can be made externally, perhaps using a spreadsheet (which is very simple to construct, incidentally).

A suitable estimator of activity standard deviation is given by:

$$\hat{s} = \frac{b - a}{6}$$

where s is the standard deviation of the expected time, t_e.

Once the critical path has been determined for the network, the standard deviation for the total critical path can be calculated by taking the square root of the sum of the variances of the activities on the critical path. Thus, in the case of only three activities on the critical path, the standard deviation would be given by:

$$\hat{s}_{cp} = \sqrt{s^2_1 + s^2_2 + \ldots + s^2_n}$$

From statistics, we know that there is a 68 percent probability of completing the project within plus or minus one standard deviation of the mean, 95 percent within two standard deviations, and 99.74 percent within three standard deviations. The normal curve is shown in Figure 11.2 for reference:

Figure 11.2. Normal Distribution.

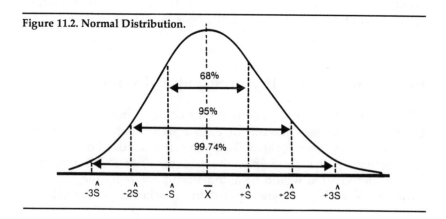

An Example

To illustrate how PERT works, we will consider a single activity, for which estimates are made by two different planners. The estimates given by each person are shown in Table 11.2, together with the calculated values for t_e and s.

Table 11.2. Estimates for a Single Activity Made by Two Individuals.

Description	Person 1	Person 2
m = most likely	10 days	10 days
a = optismistic	9 days	9 days
b = pessimistic	12 days	20 days
PERT TIME EST.	10.2 days	11.5 days
Standard Dev.	0.5 days	1.8 days

Note that the standard deviation for the estimates made by person one is only 0.5 day, meaning that the spread on the normal distribution curve is quite small. For person two, the standard deviation is 1.8 days. For convenience, we will call this 2.0 days even. The normal distribution curve, using these two different sets of numbers, would look as shown in Figure 11.3.

Figure 11.3. Distribution with Ranges.

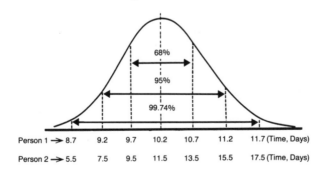

		68%			
		95%			
		99.74%			

| Person 1 → 8.7 | 9.2 | 9.7 | 10.2 | 10.7 | 11.2 | 11.7 (Time, Days) |
| Person 2 → 5.5 | 7.5 | 9.5 | 11.5 | 13.5 | 15.5 | 17.5 (Time, Days) |

The impact on the activity estimate is that the *confidence interval* for person two is four times wider than that for person one for a *given probability* of completion of the task.

To illustrate, there is a 68 percent probability that the activity will be completed in the range of 9.7 to 11.7 days if the estimates made by person one were used, whereas the 68 percent confidence interval is 9.5 to 13.5 days if the estimates made by person two were used.

What is meant by these statistics is simply that person one has greater confidence or less *uncertainty* about his estimates than person two. Does that mean he is more correct? No. It is simply a reflection of the different experiences of the two individuals.

Perhaps because person two has had less experience with this particular activity than person one, he is not sure how long it will take. Therefore, the PERT system would tell him to use an activity duration of 11.5 days as his best estimate of mean duration, whereas person one would only use 10.2 days. This can be thought of as automatically providing some "padding" for the person who has the least confidence in his estimates, although I am using the word "pad" here in a different sense than it is normally used.

Using PERT

The fact that PERT requires that three time estimates be made for each project activity and that these be plugged into formulas to calculate a time estimate and standard deviation means additional work compared to CPM. For this reason, many planners consider PERT to be not worth the effort.

Not only that, but people question the validity of the entire process. They argue that, if all three estimates are guesses, why should the weighted composite of three guesses be any better than just using the most likely estimate in the first place? Indeed, there is merit to this argument. As I see it, one principal advantage of PERT is that it makes everyone realize that durations used to specify the completion of work are not exact, but carry with them *probabilities*.

Section Four

Project Control and Evaluation

Chapter 12

Project Control and Evaluation

Control and Evaluation Principles

> Predicting the future is easy. It's trying to figure out what's going on now that's hard.
>
> —Fritz R. S. Dressler

Proper project control and evaluation are necessary if project objectives are to be met. Therefore, the design of a project control system is very important, as is the practice of proper evaluation methods. Before such systems can be designed, it is essential to understand the basic concepts and principles of control and evaluation.

Project Evaluation

> e • val • u • ate: to determine or judge the value or worth of
>
> —*The Random House Dictionary*

As the dictionary definition says, to evaluate a project is an attempt to determine if the overall status of the work is acceptable, in terms of intended value to the client once the job is finished. Project evaluation appraises the progress and performance of a job compared to what was originally planned. That evaluation provides the basis for management decisions as to how to proceed with the project. The evaluation must be credible in the eyes of everyone affected, or decisions based on that evaluation will not be considered valid. The primary tool for project evaluation is the *project audit*, which is usually conducted at major milestones throughout the life of the project. In this chapter, we will deal with the development of a sound project control system, since no audit can be successful unless proper control methods are first employed. Project auditing will be covered in the last section of this chapter.

Purposes of Project Evaluation

In Chapter 2 we saw that the last phase of a project involves a post-mortem analysis, which is conducted so that the management of projects can be improved. However, such an audit should not be conducted only at the end of the project. Rather, audits should be conducted at major milestones in the project, so that learning can take place as the job progresses. Further, if a project is getting into serious trouble, the audit should reveal such difficulty so that a decision can be made to continue or terminate the work.

I personally know of one project that was terminated when it was learned that the new product being developed was going to reach the market too late, and because of its late entry the company would never recover its investment. The cost to terminate the job was in excess of $100 million! Naturally, no organization accepts such a loss without serious soul-searching, but it was better in this case to cancel the job rather than to continue "throwing good money after bad."

Following are some of the general reasons for conducting periodic project audits:

- ☞ Improve project performance together with the management of the project.

- ☞ Ensure that quality of project work does not take a back seat to schedule and cost concerns.

- ☞ Reveal developing problems early, so that action can be taken to deal with them.

- ☞ Identify areas where other projects (current or future) should be managed differently.

- ☞ Keep client(s) informed of project status. This can also help ensure that the completed project will meet the needs of the client.

- ☞ Reaffirm the organization's commitment to the project for the benefit of project team members.

Project Control

> con • trol: to compare progress against plan
> so that corrective action can be taken when a
> deviation occurs.

Control is an attempt on a day-to-day basis to keep project work on track. It consists of measuring the status of work performed, comparing that status with what was planned to be accomplished to-date, then taking corrective action to get back on target if a deviation is discovered. The need for a good plan, against which progress can be compared, was emphasized in Chapter 4. In this chapter, the focus will be on attempting to assess or measure actual progress, which is not always such an easy task, as we shall see.

One important distinction about control should be made. The word *control* often refers to power, authority, command, or domination. Another meaning, however, is that of guiding a course of action to meet a predefined objective. This is the meaning of control that should be applied to project control systems.

Based on these ideas, here are some premises of management control systems.

☞ **Work is controlled—not workers.** The objective is to get the work done, not make workers "toe the line." Authoritarian management generally leads to resentment and an atmosphere that stifles creativity, which is just the opposite of what is needed. Control should be viewed as a *tool* that the worker can use to work more effectively and efficiently.

☞ **Control is based on completed work.** To determine if the work process is achieving objectives, the product produced is examined. In the case of a complex task, the work is subdivided (to the work-package level, for example) and the smaller units are monitored. Each task must have a well-defined output (or deliverable), and there must be standards for evaluating the completed work.

☞ **Control of complex work is based on motivation and self-control.** Control must be exercised by the person doing the work or by someone else. Control by someone other than the worker has a number of problems. Control is likely to degenerate into control of the worker, rather than the work. There is the need for communication between the worker and controller, which may not take place properly. Finally, the controller probably does not know the work as well as the worker, and cannot establish reasonable check-points as required. The worker is in the best position to establish a course of action and monitor his or her own progress.

Self-control is part of the job of every knowledgeable worker. This should be clearly spelled out to those individuals. The best set of control procedures will not work unless the people involved are motivated to make it work.

☞ **Methods of obtaining control data must be built into the work process.** That is, the person doing the work must be able to tell where he or she is at any given time. When driving, we use road signs to tell us where we are, and we compare those to our map to see if we are on course. If a brick wall is being constructed, it is easy enough to count the bricks actually laid (or measure the height of the wall) so that figure can be compared to the plan. As has been pointed out previously, however, knowledge work is harder to measure and usually will be an estimate of progress.

As a further consideration, only data that is actually required for control should be collected. The control process should not be a burden.

☞ **Control data must go to the person who does the work.** Consider a pilot. Do you give information about the plane's position to the pilot's boss? Of course not. Yet this is often done in organizations, and results in a manager's receiving more control data than he can possibly use.

161

☞ **A control system is designed for the routine.** A thermostat turns a furnace on or off to control temperature. It cannot compensate for an empty fuel tank. A control system is designed to cope with the routine: exceptions must be given special handling. It must, then, be decided what is routine and what is not.

☞ **Control of a complex process is achieved through levels of control.** That which is exceptional at one level may be routine at the next higher level. Only the most pressing problems should find their way to the top level of control.

Characteristics of a Project Control System

There are four basic activities that must be performed to have a satisfactory control system. These are as follows:

1. Planning performance

2. Observing actual performance

3. Comparing actual and planned performance

4. Adjusting as required

Comparing performance against plan can be difficult when the work cannot be quantified. When work can be quantified, deviation from plan is called *variance.* For nonquantifiable work, performance must be judged subjectively. Usually such judgment is binary—that is, the work is either satisfactory or it is not.

Summary performance reports should be standardized for all projects. Data should also be presented in an effective way. There must be a balance between presenting too much and too little data.

Objectives

The control system must focus on objectives. The designer of the control system should answer these questions:

☞ What is important to the organization?

☞ What are we attempting to do?

☞ Which aspects of the work are most important to track and control?

☞ What are the critical points in the process at which controls should be placed?

The important should be controlled. However, what is controlled tends to become important. Thus, if budgets and schedules are emphasized to the exclusion of quality, only those will be controlled. The project may well come in on time and within budget at the expense of quality.

Response

A control system should focus on response—if control data does not result in action, then the system is ineffective. That is, a control system must use deviation data to *initiate corrective action* or it is not really a control system but simply a monitoring system.

Timeliness

The response to control data must be timely. If action occurs too late, it will be ineffective.

Human Factors

The system should be easy to use. In particular, the control system should be designed for the convenience of people and not machines.

Flexibility

One system is not likely to be correct for all projects. It may need to be scaled down for small projects and beefed-up for large ones.

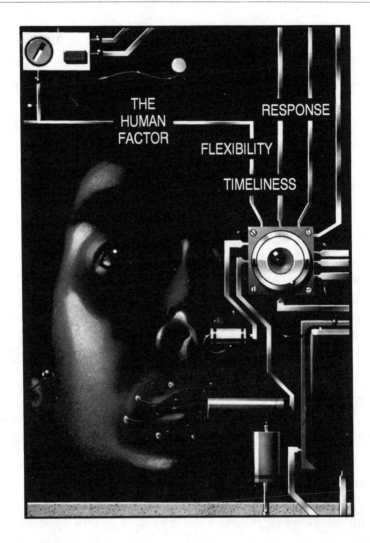

Simplicity

The smallest control effort that achieves the desired result should be used. Any control data that is not essential should be eliminated. However, one common mistake is to try to control complex projects with systems which are *too simple!*

Components of a Project Control System

In its simplest form, a project control system can be represented by a first-order feedback system, as shown in Figure 12.1. The system has *inputs*, *outputs*, and a *process* for transforming those inputs to outputs, together with a *feedback loop* to ensure that the system continues processing inputs according to its design. The outputs are monitored, compared to some pre-set standard, and if the outputs are not correct, that information is fed back as an input to the system to correct for the deviation.

Figure 12.1. First-Order Feedback System.

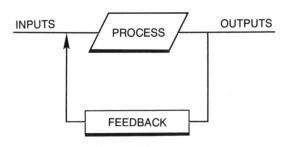

This feedback system is a very simple one. It is not very elegant, and has some serious limitations as a model of how to achieve control in project management.

For those readers unfamiliar with feedback systems, a good analogy for the first-order system is the thermostat in one's home. In the winter, the system provides heat and the desired room temperature is preset by adjusting the thermostat to the proper level.

It should be clear that every system is designed to work properly only under certain conditions. For example, the home heating system might be designed to maintain the room at 70 degrees Fahrenheit so long as the outside temperature does not go below minus 30 degrees. When the outside temperature drops below that level, the heater will run continuously, but the room temperature will begin to drop below the preset level of 70 degrees.

To maintain the desired room temperature, the system would have to increase its heating capacity, but it cannot do this. Thus, it keeps running, without being able to adequately heat the house.

In a similar manner, a project may run into unexpected obstacles, which fall outside the boundaries for which the project control system was designed. Everyone is following the plan to the letter, but they are not getting the desired result. What is needed is to change the approach. However, a first-order control system does not have that capability. Something more flexible is needed. The third-order system shown in Figure 12.2 is the answer.

Figure 12.2. Third-Order Feedback System.

The Project Control System (adapted from Burrill & Ellsworth; see reading list)

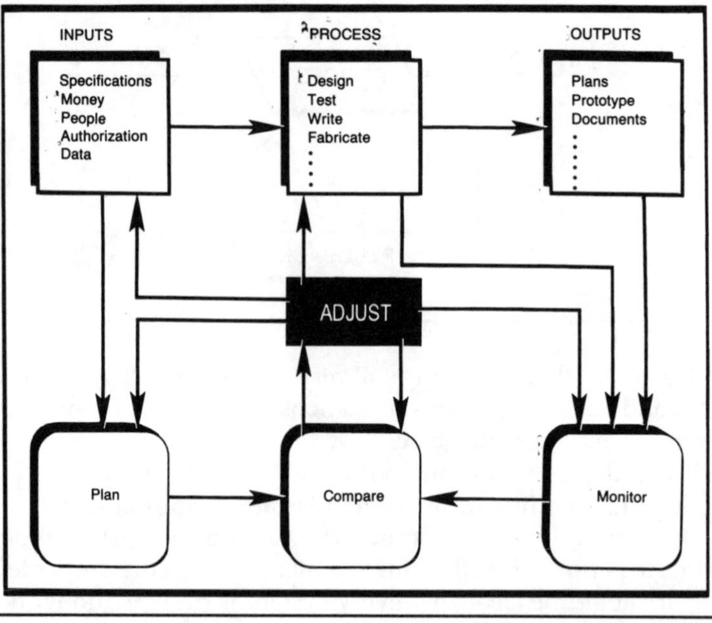

The system in Figure 12.2 has the same basic elements as the first-order system of Figure 12.1. There are *inputs, processes, outputs,* and *feedback.* However, the third-order system feeds information about the system outputs to a *comparator,* which weighs them against the original plan. If there is a discrepancy, that information is passed to an *adjust* element, which must decide if the discrepancy is caused by something being wrong with the process, the inputs, or the plan itself.

Once that determination is made, the adjust element calls for a change in the plan, inputs, or the process itself. Note also that the adjust element has an arrow going back to the monitor. If a deviation is detected, the monitoring rate is increased until the deviation is corrected, then monitoring is decreased to its original level.

The real-world analogy is that if you were monitoring progress on a project weekly and a problem occurred, you might begin to monitor daily. If the problem becomes serious enough, your monitoring rate might increase to several times each day. Once the problem has been solved, you would revert to your weekly monitoring.

Comparing performance against plan can be difficult when the work cannot be quantified. How do you know what percentage of a design is complete, for example? Or if you are doing a mechanical drawing of a part, is the drawing 75 percent complete when 75 percent of the area of the paper is covered? Probably not. Measuring progress in *knowledge work*, to use Peter Drucker's term, is very difficult.

This often leads to strange results. Suppose a member of the project team has agreed to design a new golf club, and has promised to finish it in ten weeks. At the end of week one, she reports that the design work is 10 percent complete. At the end of week two, the work is 25 percent complete. In week three she hits a small snag and gets a little behind, but by week five she has gotten ahead again. Figure 12.3 shows a plot of her progress.

Everything goes pretty well until week eight, when she hits another snag. At the end of that week, she has made almost no progress at all. The same is true the following week, and the following, and the following . . .

What happened? For one thing, the 80/20 rule got her. In the case of knowledge work, it says that 80 percent of the work will be consumed by 20 percent of the problems encountered, and they will always happen near the end of the job.

The real issue, though, is how she measured progress in the first place. Chances are, at the end of the first week, she reasoned somewhat like Cathy, of comic strip fame who said to herself,

Figure 12.3. Percent Complete Report.

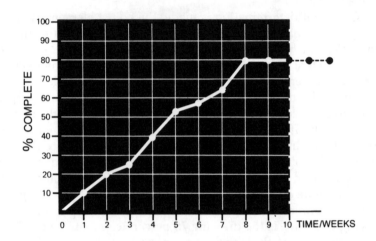

"I'm at the end of the first week on a ten-week job. I must be 10 percent complete." And she would be in good company, because that is exactly what a lot of people do when *estimating* progress on knowledge work.

> I still have checks—I must have money in the bank!
> —Cathy

Note the word *estimate!* Assessing progress when work is not easily quantifiable is estimating, and subject to all the difficulties discussed in Chapter 7. This shows the limits of our ability to achieve control in management.

It is for this reason that two practices are advisable. First, work should be broken into small "chunks" that permit progress to be monitored fairly frequently, perhaps at intervals no greater than two weeks. Second, tangible deliverables should be used as signposts to show progress. In design, a drawing is tangible evidence of progress. The same is true with software development. Printed code or written functional specs are evidence that work is complete. Having it "in one's head" is impossible to verify.

Conducting the Project Audit

Ideally, a project audit should be conducted by an independent examiner, who can remain objective in the assessment of information. However, the audit must be conducted in a spirit of learning, rather than in a climate of blame and punishment. If people are afraid that they will be "strung up" for problems, then they will hide those problems if at all possible.

Even so, this is hard to achieve. In many organizations the climate has been punitive for so long that people are reluctant to reveal any less-than-perfect aspects of project performance. Chris Argyris (1990) has described the processes by which organizations continue ineffective practices. All of them are intended to help individuals "save face" or avoid embarrassment. In the end, they also prevent organizational learning.

To say it simply, there is no such thing as *constructive criticism* when you are the target of it. Criticism—even in the form of "we could have done this better"—hurts, and we generally avoid being hurt if possible. For that reason, it takes organizations time and considerable effort to develop a climate in which project auditing can be done in a true spirit of cooperation and a desire to learn and improve.

As a sequel to this, it seems clear that auditors must be chosen carefully. An auditor with a NIGYSOB (Now-I've-Gotcha-You-Son-Ova-Bitch) mentality is certain to create more problems than solutions.

The Audit Report

There may be varying degrees of audits conducted, from totally comprehensive, to partial, to less formal, cursory examinations. A formal, comprehensive audit should be followed by a report, which should contain as a minimum the following:

1. **Current project status.** This is best shown using earned-value analysis, as presented in Chapter 13. However, when earned-value analysis is not used, status should still be reported with as much accuracy as possible.

2. **Future status.** This is a forecast of what is expected to happen in the project. Are significant deviations expected in schedule, cost, performance, or scope? If so, the nature of such changes should be specified.

3. **Status of critical tasks.** The status of critical tasks, particularly those on the critical path, should be reported. Tasks that have high levels of technical risk should be given special attention, as should those being performed by outside vendors or subcontractors, over which the project manager may have limited control.

4. **Risk assessment.** Have any risks been identified that highlight potential for monetary loss, project failure, or other liabilities?

5. **Information relevant to other projects.** What has been learned from this audit that can and should be applied to other projects, whether presently in progress or about to start?

6. **Limitations of the audit.** What factors might limit the validity of the audit? Are any assumptions suspect? Is any data missing or suspect of contamination? Was anyone uncooperative in providing information for the audit?

As a general comment, the simpler and more straightforward a project audit report, the better. The information should be organized so that planned versus actual results can easily be compared. Significant deviations should be highlighted and explained. In Figures 12.4 and 12.5 are forms intended to be used for very simple project audits. For more comprehensive ones, the forms will not go far enough. See also the project checklists in Chapter 20 for additional ideas.

Figure 12.4. Project Audit or Post-Mortem Analysis Form.

Project:	
Prepared by:	**Date:**
For the period from	**to**
Evaluate the following objectives:	
Performance was ○on target ○above target ○below	
Budget was ○on target ○overspent ○underspent	
Schedule was ○on target ○behind ○ahead	
Overall, was the project a success? Yes No	
If not, what factors contributed to a negative evaluation?	
What was done really well?	
What could have been done better?	
What recommendations would you make for future project applications?	
What would you do differently if you could do it over?	
What have you learned that can be applied to future projects?	

Figure 12.5. Project Reporting Form.

Project:	
Prepared by:	Date:
For the period from	to
Accomplishments for report period are:	

We are /◯on schedule /◯ahead /◯behind

List any change to project objectives:	List any changes which we have experienced in business climate:
What unanticipated problems do we face?	
What changes are needed?	List anyone whose approval is needed for changes:
Action steps which I plan to take:	List any additional unanticipated problems:
Comments:	

Chapter 13

Project Control Using Earned-Value Analysis*

Using Variance or Earned-Value Analysis in Project Control

Even though there are limits in assessing exactly how much work has been done on a project, there can be no control unless some

---

*   Much of this material is adapted from my book *Project Planning, Scheduling & Control (Chicago: Probus, 1991)*.

assessment is done. The most widely used method of measuring project progress is through *variance* or *earned-value* analysis. Earned-value analysis is the heart of the Cost/Schedule Control Systems Criteria (C/SCSC) developed in 1963 by the Department of Defense. For a more detailed explanation of the C/SCSC system, see Chapter 18 on Progress Payments. What follows in this chapter is an abbreviated approach, which should be appropriate for application to most projects, not just large government jobs. For those readers who need a complete treatment of the C/SCSC system, see Fleming, *Cost/Schedule Control Systems Criteria: The Management Guide to C/SCSC*, Revised Edition (Chicago: Probus, 1992).

First, we define variance as follows:

> Variance: any deviation from plan.

Variance analysis allows the project manager to determine "trouble spots" in the project and to take corrective action. As was mentioned previously, there are three areas of the project that the project manager is expected to control. These are the *performance, cost, time* objectives. Because of the difficulty of quantifying the performance objective, variance analysis is usually applied only to the cost and schedule targets. For that reason, the project manager will have to monitor the quality targets, using whatever standards can be developed, and take necessary steps to ensure that they are met.

As for the schedule and cost objectives, the following terms define what is to be monitored:

☞ **Cost Variance:** Compares deviations only from budget and provides no comparisons of work scheduled and work accomplished.

☞ **Schedule Variance:** Compares planned versus actual work completed. This variance can be translated into the dollar value of the work, so that all variances can be specified in monetary terms.

> In some organizations, project managers do not deal
> with costs, but rather with labor hours. Once standard
> variance analysis has been presented in terms of cost,
> a method of dealing with working hours will be pre-
> sented.

In order to make cost and schedule variance measurements,
three variables are used. They are defined in the following para-
graphs, together with examples of how they are calculated.

☞ **BCWS** (Budgeted Cost of Work Scheduled): The budgeted
cost of work scheduled to be done in a given time period,
or the level of effort budgeted to be performed in that period. This is the *target* toward which the project is headed.

Another way to say it is that BCWS represents the *plan* which
one is supposed to follow. It is basically the product of man-hours
and the dollar labor rate that is paid during a given period of time,
usually a day or week at a time.

As an example, suppose that a project is to employ two people
working on the project for one week (40 hours) at the labor rate of
$30 per hour each (loaded labor—with overhead included). In addi-
tion, a third person will work on the project for 30 hours during the
same week, but at a loaded labor rate of $50 per hour. The budg-
eted cost of work scheduled for the week, then, is the sum of two
products:

$$40 \text{ hours} \times 30 \text{ dollars/hour} \times 2 = \$2,400$$
$$30 \text{ hours} \times 50 \text{ dollars/hour} \times 1 = \$1,500$$
$$\text{Total BCWS} = 2,400 + 1,500 = \$3,900$$

☞ **BCWP** (Budgeted Cost of Work Performed): The budgeted
cost of work actually performed in a given period. BCWP
is also called **earned value.** It is a measure of how much
work has been accomplished.

The BCWP figure is calculated as follows. For the example above, assume that the two employees who are assigned to work for a full 40 hours each do indeed put in that amount of effort. One worker actually gets her work complete, while the other does not. He only completes about 80 percent of the work supposed to be done. The worker assigned to put in only 30 hours also completes his work as planned. We say that the *earned value* of the work completed, then, is as follows:

$$
\begin{aligned}
40 \text{ hours} \times 30 \ \$/\text{hour} &= \$1,200 \\
0.8 \times 40 \text{ hours} \times 30 \ \$/\text{hour} &= \$ \ 960 \\
30 \text{ hours} \times 50 \ \$/\text{hour} &= \underline{\$1,500} \\
\text{BCWP TOTAL:} &= \$3,660
\end{aligned}
$$

☞ **ACWP** (Actual Cost of Work Performed): The amount of money actually spent in completing work in a given period. This is the amount of money paid to workers (wages only—no material costs are included in any of these figures) to do the work that was completed during the time period in question.

To continue with the above example, assume that the work completed has actually cost the organization $3,900. If this figure were compared with BCWS, we might think the project is in good shape. The scheduled work was supposed to cost $3,900, and that is what has been paid in labor. However, we also know that one person did not get through with the work he was supposed to do. The value of his accomplishment is only $960, but was supposed to be $1,200. In order to see what this means for the project, the following formulas are employed:

Cost Variance = BCWP – ACWP
Schedule Variance = BCWP – BCWS (Dollar value)

Plugging numbers into these formulas, we have the following results:

Cost Variance = $3,660 – $3,900 = –$240

A negative cost variance means that the project is spending more than it should—thus, a negative variance is *unfavorable*.

Schedule Variance = $3,660 – $3,900 = –$240

Again, a negative schedule variance means that the project is behind schedule, and so is also *unfavorable*.

Looking at these two figures together tells us that the project has gotten behind schedule in the amount of $240 worth of work, and since the cost variance is identical to the schedule variance, we know that the cost variance is due *only* to the schedule variance. That is, the work being done is costing what it was estimated to cost. If labor rates had escalated, then the cost variance would be greater than the schedule variance.

Variance Analysis Using Spending Curves

Variances are often plotted using spending curves. In Figure 13.1 is a BCWS curve for a project. It shows the *cumulative spending* planned for a project, and is sometimes called a *baseline plan*. Such curves can often be plotted automatically by transferring spending data from a scheduling program (which calculates labor expenses on a daily or weekly basis by multiplying labor rates times manpower expended) to a graphics program using a DIF file or some other file-transfer format.

In the event that software is not available to provide the necessary data, Figure 13.2 shows how data for the curve is generated. Consider a simple bar-chart schedule. Only three tasks are involved. Task one involves 40 labor-hours per week at an average loaded labor rate of $20/hour, so that task spends $800/week. Task two involves 100 hours/week of labor at $30/hr, so it costs $3,000/week. Finally, task three spends $2,400/wk, based on 60 hrs/wk of labor at $40/hr.

At the bottom of the chart we see that during the first week $800 is spent for project labor; in the second week both tasks one and two are running, so the labor expenditure is $3,800. In the third week, all three tasks are running, so labor expenditure is the sum of the three, or $6,200. These are the *weekly* expenditures.

Figure 13.1. A Cumulative Spending or BCWS Curve.

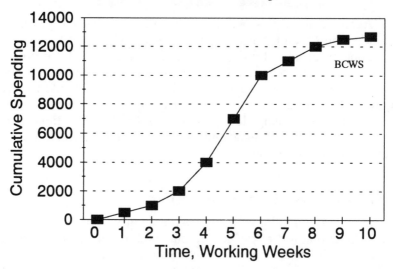

The *cumulative* expenditures are calculated by just adding the cost for each subsequent week to the previous cumulative total. At the end of week one, $800 has been spent. At the end of week two, the figure is $4,600; at week three, it is $10,800, and so on.

These cumulative amounts are plotted in Figure 13.3. This is the spending curve for the project, and is called a BCWS curve. Since it is derived directly from the schedule, it represents *planned performance,* and therefore is called a *baseline plan.* Further, since control is exercised by comparing progress to plan, this curve can be used as the basis for such comparisons so that the project manager can tell the status of the program. Following are examples of how such assessments are made.

Figure 13.2. Calculation of Cumulative Labor Spending from a Bar Chart.

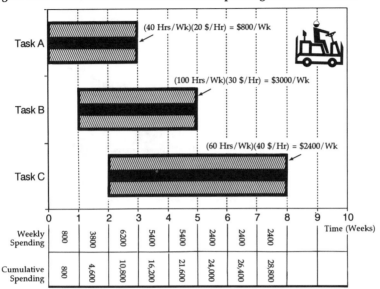

	0	1	2	3	4	5	6	7	8	9	10
											Time (Weeks)
Weekly Spending		800	3800	6200	5400	5400	2400	2400	2400		
Cumulative Spending		800	4,600	10,800	16,200	21,600	24,000	26,400	28,800		

Examples of Progress Tracking Using Spending Curves

Consider the curves in Figure 13.4. On a given date, the project is supposed to have involved $50,000 (50K) in labor (BCWS). The actual cost of the work performed (ACWP) is $60,000. These figures are usually obtained from Accounting, and are derived from all of the time cards that have reported labor applied to the project. Finally, the budgeted cost of work performed (BCWP) is $40,000. Under these conditions, the project would be behind schedule and underspent.

To understand this, the project has spent $60,000 to accomplish only $40,000 worth of work. That means an overspend of $20,000. The plan called for $50,000 worth of work to be done, but they have only completed $40,000 worth, so they are behind schedule as well.

Figure 13.5 illustrates another scenario. The BCWP and ACWP curves both fall at the same point, $60,000. This means that the project is ahead of schedule, but spending correctly for the amount of work done. To see this, the project has spent $60,000 and has ac-

Figure 13.3. Cumulative Spending Curve for the Bar Chart in Figure 13.2.

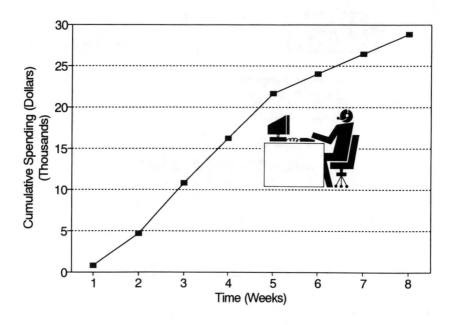

complished $60,000 worth of work. However, the plan called for $50,000 worth of work, so the status is ahead of schedule.

Is there any potential problem with being in this position? At first glance, there is none. However, if you consider how a project can be in this position, you find that more resources must have been applied than were planned, but at the planned labor rate (since there is no spending variance). Then the question is, where did the project manager get the extra people? In most environments, resources are shared, so it may be that this project is ahead of schedule at someone else's expense.

Another consideration is cash flow. While the project is ahead of schedule, can it be funded at the rate being spent? If not, then the work would have to be decelerated.

Figure 13.4. Project Behind Schedule and Overspent.

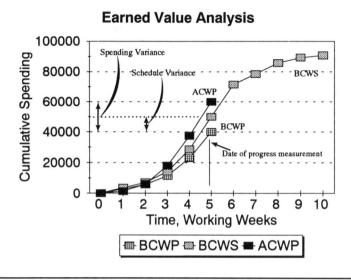

Figure 13.5. Project Ahead of Schedule and Above Budget.

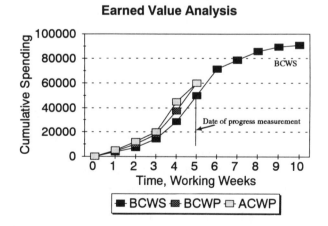

The next set of curves illustrates another status. In Figure 13.6 the BCWP and ACWP curves are both at $40,000. This means the project is behind schedule and under budget. This project is probably starved for resources (the victim of another project manager being ahead). Labor is costing what it is supposed to cost, but not enough work is being done to stay on schedule. The problem for this project manager is that she will probably go over budget in trying to catch up, since premium labor will most likely be required.

Figure 13.6. Project Behind Schedule and Below Budget.

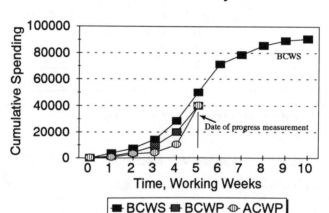

Finally, Figure 13.7 looks like Figure 13.4, except the ACWP and BCWP curves have been reversed. Now the project is ahead of schedule and underspent. The accomplished work has an earned value of $60,000, but the actual cost of that labor has been only $40,000. There are three possibilities that can explain how this project manager achieved the result shown.

Figure 13.7. Project Ahead of Schedule, Underspent.

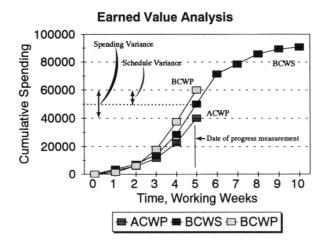

1. Actual labor rates were considerably lower than expected, and the people were more efficient than anticipated.

2. The project team had a "lucky break." They had expected to have to work really hard to solve a problem, but it turned out to be very easy.

3. The project manager "sand-bagged" his estimates. He padded everything, playing it safe.

If you believe situation one, you will believe anything. It is very unlikely that both variances would happen at the same time.

Situation two happens occasionally. When all the planets are aligned—about once in a zillion years, you say. You bet!

Situation three is the most likely explanation. The project manager was playing it safe. And he would tell you that there is no problem. After all, the project will come in slightly ahead of schedule and underspent, which means he will give money back to the company. No problem. The controller no doubt has the budgeted funds in an interest-bearing account, so the company is earning interest on the money.

Right. But banks are not known for being overly generous with interest, and there is a rule that says that if you can't make a greater return on an investment than the interest a bank will pay, you should go out of business and just put your money in the bank. You don't need the aggravation of being in business, and you aren't very good at it anyway.

As a matter of fact, the economists would say that there is a real *opportunity cost* involved in this project. The company has lost an opportunity to get a good return on its investment because the money was budgeted for this project, and therefore could not be used anywhere else. This is another problem with padding.

The question is, naturally, what is reasonable? We certainly cannot expect to have zero variances in a project. And this is true.

However, there is no easy answer to the question. Well-defined construction projects can be held to very small tolerances—as small as plus-or-minus three to five percent. Research and development projects are likely to run higher tolerances, perhaps in the range of 15 to 25 percent. Each organization has to develop acceptable tolerances based on experience.

Refining the Analysis

The only problem with the analysis presented here is that it is an *aggregate* figure, and would not permit determination of what area of the project a problem exists in, and may even hide a problem completely. For that reason, the variance analysis needs to be conducted on a task-by-task basis.

This is usually done at the Work Package level, but it can be at any level of the Work Breakdown Structure at which one wishes to track project progress. By summing the individual Work Package figures, the aggregate figure can be used to gauge the overall "health" of the project, while a line-by-line accounting can be used to spot specific problem areas.

The importance of this was brought home to me by a client who reported that they had been using aggregate analysis to gauge project status for some time, and they discovered that a $100,000 overspend in one area of a project was being counter-balanced by a $100,000 underspend in another area. It looked like the project was in good shape, but such huge variances indicate a lack of control, and should be addressed.

Variance thresholds can be established that define the level at which reports must be sent to various levels of management within an organization.

By combining cost and schedule variances, an integrated cost/schedule reporting system can be developed. In Figure 13.8 is a form which illustrates the use of this concept to track a project. The form has been filled in with some data to illustrate the various combinations of the numbers and their meanings.

The Project Status Report shows the levels of project costs and work completed to-date for each Work Package (or whatever level you wish to use to report progress). The report is configured as a QuatroPro® spreadsheet. The columns contain the following information:

☞ Column 1: The Work Package number

☞ Column 2: BCWS (Budgeted Cost of Work Scheduled to-date). Referring back to Figure 13.2, for task 1, at the end of the first week, the BCWS figure is $800. At the end of the second week, it is $1600. Note that for task 2, nothing will be entered into this cell until week 2.

☞ Column 3: BCWP (Budgeted Cost of Work Performed to-date). This is the *earned value* figure defined previously.

☞ Column 4: ACWP (Actual Cost of Work Performed to-date). This is the actual cost of labor to-date, as previously defined.

☞ Column 5: Schedule Variance—the difference between BCWS and BCWP—calculated by the spreadsheet.

Figure 13.8. Project Status Report Using Earned-Value Analysis.

Project Status Report

Project No.:	201				Date:	07-Dec-92		FILE:	PROJRPT1	
Description:	Communications Receiver				Page	___ of ___				
Prepared by:	James P. Lewis				Signed:					

WBS #	Cumulative-to-date			Variance		At Completion			Critical	Action
	BCWS	BCWP	ACWP	Sched.	Cost	Budgeted	Latest Est.	Variance	Ratio	Required
501	12,400	12,400	12,600	(300)	(200)	15,750	15,750	0	0.98	O.K.
502	7,200	7,700	7,700	500	0	9,000	8,500	500	1.07	O.K.
503	16,200	14,000	14,000	(2,200)	0	21,000	22,000	(1,000)	0.86	CHECK
504	3,100	2,000	1,900	(1,100)	100	5,000	5,000	0	0.68	RED FLAG
505		4,750	4,750	0	0	4,750	4,750	0	NA	NA
506	6,500	7,500	5,650	1,000	1,850	8,000	8,000	0	1.53	RED FLAG
507	500	500	600	0	(100)	2,000	2,400	(400)	0.83	CHECK
508				0	0	18,000	18,000	0	NA	NA
509				0	0	12,000	12,000	0	NA	NA
510	2,200	500	500	(1,700)	0	4,200	5,000	(800)	0.23	RED FLAG
				0	0			0	NA	NA
				0	0			0	NA	NA
				0	0			0	NA	NA
TOTAL	48,200	44,600	42,950	(3,600)	1,650	90,700	*****	(1,700)	0.96	O.K.

NOTE: Negative variance is unfavorable ‖ If Critical Ratio < 0.6, INFORM MANAGEMENT!

( ) = NEGATIVE VALUES

☞ Column 6: Budget Variance—the difference between ACWP and BCWP—calculated by the spreadsheet

☞ Column 7: The at-completion target cost for the work. For Figure 13.2, task 1, the at-completion cost for labor will be $2,400 (three weeks at $800/week). For task 3, the at-completion cost will be $14,400 (six weeks at $2,400/week). Naturally, labor spending will not always be uniform. This example uses uniform spending for simplicity.

☞ Column 8: The latest estimate of what the work will cost when complete. If we find on task 1 that we are actually having to spend $22 per hour for labor, rather than the $20 that was originally budgeted, but we expect the work to be completed in the same number of working hours that was originally estimated, then the budget-at-completion (BAC) will be 22 × 40 × 3 or $2,640, rather than the originally budgeted $2,400, or an overspend of $240. It is also possible that the BAC can differ from the original estimate because more or less labor hours will be needed, but labor costs will be what were originally planned. The BAC figure is extremely important when decisions are being made as to whether to continue or terminate a project.

☞ Column 9: The at-completion budget variance expected—difference between columns 7 and 8, calculated by the spreadsheet

☞ Column 10: The critical ratio—calculated as described below

☞ Column 11: Action required—determined by the spreadsheet using an "IF-formula." These rules are explained below.

The Critical Ratio

Part of the C/SCSC system involves calculation of two ratios that indicate how well the project is doing. One of these is called a *Cost*

Performance Index (CPI) and the other is called a *Schedule Performance Index* (SPI). The CPI is the ratio of BCWP to ACWP, or (BCWP/ACWP). The SPI is the ratio of BCWP to BCWS, or (BCWP/BCWS). Meredith & Mantel (1985) describe a control-charting method that can be used to analyze progress in projects. They calculate a *critical ratio*, which is the product of the CPI and SPI, using the following formula:

$$\text{Critical Ratio} = (\text{CPI}) * (\text{SPI})$$

or

$$\text{Critical Ratio} = \frac{\text{BCWP}}{\text{BCWS}} * \frac{\text{BCWP}}{\text{ACWP}}$$

As is true for control charts used to monitor manufacturing processes, rules can be devised for responding to the critical ratio. Meredith and Mantel (1985) suggest limits and actions as shown in the diagram in Figure 13.9. These limits are only suggestions, and the project manager will have to devise limits that are appropriate for his or her own programs.

Using a spreadsheet allows the process of interpretation to be automated. The progress report in Figure 13.8. is set up with an "IF" formula in the final column. This formula looks at the Critical Ratio calculated in the previous column and subjects it to tests. Based on those tests, the formula returns the words "O.K.," "CHECK," "RED FLAG," or "NA," meaning no critical ratio has yet been calculated in the cell being tested. The tests are simple. In words:

☞ Print "O.K." if the critical ratio (CR) is between the values 0.9 and 1.2

☞ Print "CHECK" if the CR is between 0.8 to 0.9 or 1.2 to 1.3

☞ Print "RED FLAG" if the CR is above 1.3 or below 0.8

In addition, if the ratio falls below 0.6, company management should be informed, as progress is so much better than expected that some changes probably should be made to the project plan.

Figure 13.9. Critical Ratio Control Limits.

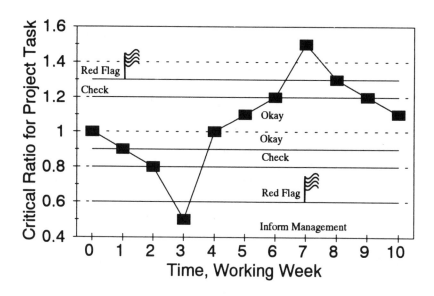

Following is the IF-formula for a spreadsheet with the critical ratio in cell K10 and the IF-formula in cell L10.[1]

@IF(K10>1.3#OR#K10< 0.8, "RED FLAG", @IF(K10>1.2#AND#
K10<1.3#OR#K10>0.8#AND#K10<0.9, "CHECK","O.K."))

Note that the example shown is set up so that the bottom-line summary for the project looks very good. The critical ratio for the overall project is 0.96, indicating that everything is fine. However, there are three work packages with RED FLAGS and two with CHECKS indicated, which means that some parts of the project are in trouble. If this were my project, I would be concerned.

However, a complete assessment of the project cannot be done using just this report. We also need to know exactly where in the project these work packages fall. Are any of the ones with problems on the critical path? If so, we know we have a more serious problem than indicated by the summary analysis. Even if none are critical, are any of them behind far enough to be running out of float? If so, then they will soon be critical. (See, for example, work packages numbered 504 and 510. These are far enough behind that they could be in real trouble.)

The Need for All Three Measures

Occasionally project managers fall into the trap of trying to track their projects using only BCWS and ACWP. As long as they see no difference between what they had planned to spend and what has actually been spent, they think the project is running smoothly. However, we saw from the above examples that this may not be true, and the manager would not spot a problem until it had perhaps gotten serious.

In fact, a controller from one organization told me that he constantly sees this happen in his company. For a long time the project goes along being underspent or right on target. Then the project manager realizes that the work is not getting done as required, and a big effort is applied to catch up. The usual result is that spending overshoots the planned target. This is illustrated by the curves in Figure 13.10.

Variance Analysis Using Hours Only

In some organizations, project managers are not held accountable for costs, but only for the hours actually worked on the project and for the work actually accomplished. The argument used to justify this way of working is that project managers usually have no control over labor rates. This is because an individual may be assigned to the project by his functional manager simply because he was the only person available who could do the job, but his rate is 25 percent higher than what the project manager expected to pay when original estimating was done.

The other cause of problems is that the accounting department may change burden allocation rates (for valid reasons), which

Figure 13.10. Typical Result of Tracking Only BCWS and ACWP.

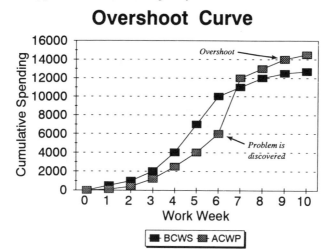

causes total labor costs to go above original estimates. This naturally creates a cost variance in the project, but one over which the project manager had no control, and so the argument is that he should not be held accountable.

In this case, the same analysis can be conducted by stripping the dollars off the figures. This results in the following:

☞ BCWS becomes Total Planned (or Scheduled) Hours

☞ BCWP becomes Earned Hours (Scheduled hours × percent work accomplished)

☞ ACWP becomes Actual Hours Worked

Using the new numbers, it is possible to compute the following variances:

☞ Schedule Variance = BCWP – BCWS = Earned Hours – Planned Hours

☞ Labor Variance = BCWP – ACWP = Earned Hours – Actual Hours Worked

An Example Using Project Workbench™

To illustrate project tracking with software, we will use the schedule developed in Chapter 10 to show resource allocation. For your convenience, the schedule is repeated in Figure 13.11. Assume that we want to show status at the end of the first week. The week ends on June 21, so we can set the report date accordingly.

Figure 13.11. ZEROMM Schedule in Original Form.

```
―――――――――――――――――― Project description ――――――――――
 Report: Gantt Chart            Date: 12-07-92            Time: 15:29

 Title: Project Zero            Project ID: SP            Version: 1

 Manager: Jim Lewis             Project filename: ZEROMM1  Dept:

 Project start:  6-17-91        Project end:             Budget:

 Description:  Small Network to illustrate resource scheduling
―――――――――――――――――――――― Legend ――――――――――――――――
 ■■■■■  Activity                    ◆         Milestone
 C■■■■  Activity on critical path   L■■■■     Locked activity
 ═■■■   Partially completed activity ═════    Completed activity
 ▲      Original start date         ▼         Original end date
 .....  Discontinuous activity      .....◆    Baselined milestone
```

Project Zero	Day	Resrc	10	17	24	1	8	July 1991 15	22	29	5	August 12
Activity A	8	TT	.	■■■■■■■■	.	.	.	.	.	.	.	.
Activity B	18	SS CC	.	C■■■■■■■■■■	■■	.	.	.	.	.	.	
Activity C	3	CC	.	.	■■■	.	.	.	.	.	.	.
Activity D	9	TT CC	.	.	.	■■■■■■	■■	.	.	.	.	.
Activity E	5	MM	.	.	.	.	■■■■■■	.	.	.	.	.
Activity F	3	MM	.	.	.	.	.	■■■	.	.	.	.
Activity G	30	TT SS	.	.	.	.	C■■■■■■■■■		.	.	.	.
Finish	◆	X	.	.	.	.	.	C◆	.	.	.	.

Project Zero	Day	Resrc	10	17	24	1	8	July 1991 15	22	29	5	August 12

Resource summary
Utilization

							July 1991		
Tom Trump	5.0	TT	5.0	5.0	4.0	5.0	5.0	1.0	
Sue Simms	5.0	SS	5.0	5.0	4.0	5.0	5.0	1.0	
Charlie Cla	5.0	CC	3.0	2.0	2.5	4.5	5.0	1.0	
Mary Martin	5.0	MM				4.0	4.0		
UNASSIGNED		X							

Only two tasks are scheduled to have any work done on them during this period. For task A, Tom Trump was scheduled to do five days of work on an eight-day-duration activity. Let's say that Tom has gotten behind. Instead of having all five days of work completed, he has only 40 percent of the total done. That means he has done a little more than what should have been done in three days. He estimates that it will take almost all of the remaining five days to complete the job. However, he has actually put in a total of four days' effort on the job, meaning that his labor costs should be about half of the total originally planned for the job.

For task B, Charlie Clark was supposed to do three days of work up-front, simultaneously with five days of work for Sue Simms being required. We will say that they have done exactly what was planned and are right on target. The total duration for activity B was originally planned to be 15 days. We will say that what was planned in the first five days has been accomplished. Does that mean that the work is 33 percent complete?

For this task the answer is no. The work is not linear. More work per day is done in the first three days than in the remaining 12 days. This is where earned value analysis provides a good measuring method. From the original plan, we find that the loaded labor rate for Charlie Clark is $160 per day and for Sue Simms it is $240. This means that the BCWS for the first five days of the activity is:

$$\$160 \times 3 = \$480 \text{ for Charlie Clark}$$
$$\$240 \times 5 = \$1,200 \text{ for Sue Simms}$$
$$\text{TOTAL} = \$1,680 = \text{BCWP}$$

The total expenditure for the activity would be ten more days of work by Sue Simms, added to the total for the first five days, or:

$$\text{BAC} = \$1,680 + 10\,(240) = \$4,080.$$

That is, the total planned cost for activity B is $4,080. If everything is right on target at the end of the first five days, then the amount of work completed is the ratio of the two figures or:

$$\% \text{ Complete} = \text{BCWP/BAC} = 1680/4080 = 41\%.$$

The question is, how do you show this on a schedule diagram? For a 15-day task, if you show 41 percent complete, it would look like six days of work are finished, which is not the case in a times-cale sense. There are still 10 days of work remaining for Sue to do. This illustrates the difficulty of showing progress on bar charts. Only earned-value analysis provides a viable way of presenting status. In the case of Project Workbench™, if you choose to display percent complete, you have the schedule shown in Figure 13.12, which indicates that 41 percent of activity B is finished, and is a bit misleading. The only way in which progress reporting on a bar

Figure 13.12. Sample Progress Report.

```
                        ─── Project description ───
  Report: Gantt Chart            Date:  6-21-91              Time: 17:48

  Title: Project Zero            Project ID: SP              Version: 1

  Manager: Jim Lewis             Project filename: ZEROMMT2 Dept:

  Project start:  6-17-91        Project end:                Budget:

  Description:  Project tracking with earned-value analysis
                              ─── Legend ───
  ■■■■■  Activity                      ◆      Milestone
  C■■■■  Activity on critical path    L■■■■   Locked activity
  ═■■■   Partially completed activity ═══     Completed activity
  ▲      Original start date          ▼       Original end date
  .....  Discontinuous activity       .....◆  Baselined milestone
```

			──July 1991──					──August 1991──	
Project Zero	Day	Resrc	17	24	1	8	15	22	29	5	12	19
Activity A	9	TT	═■■■■■									
Activity B	18	SS CC	C═════■■■■■ ■■									
Activity C	3	CC		■■■								
Activity D	9	TT CC		■■■■■ ■■								
Activity E	5	MM			■■■■■■							
Activity F	3	MM			■■■							
Activity G	30	TT SS			·C■■■■■■■■■						
Finish	◆	X				C◆					

			──July 1991──					──August 1991──	
Project Zero	Day	Resrc	17	24	1	8	15	22	29	5	12	19
Resource summary Utilization												
Tom Trump	5.0	TT	4.0	7.0	4.0	5.0	5.0	1.0			
Sue Simms	5.0	SS	5.0	5.0	4.0	5.0	5.0	1.0			
Charlie Cla	5.0	CC	3.0	2.0	2.5	4.5	5.0	1.0			
Mary Martin	5.0	MM	"			4.0	4.0				
UNASSIGNED		X									

graph can be shown unambiguously would be to separate the tasks being performed by each person so that you can report on them individually, and even this method is not perfect, since most work is a bit nonlinear.

However, an earned value analysis is very clear in showing status. An earned-value report printed from Workbench™ is shown in Figure 13.13.

The report shows that activity A is behind schedule and overspent already, and the EAC shows that the task is expected to cost $2,160, compared to the original $1,920 figure, or an overspend of $240, which is the extra day of work that Tom expects to put into the activity in order to complete it. Sue is not forecasting an at-completion variance, so the total project variance is $16,720 minus $16,480, or $240.

Endnote

[1] A floppy disk is available from the author containing many of the forms in this handbook in Quatro Pro™ format (DOS only). The cost of the disk is $25 postpaid. Contact Jim Lewis, 302 Chestnut Mountain Drive, Vinton, VA 24179; phone 703-890-1560.

Figure 13.13. Project Status Using Earned-Value Analysis from Project Workbench.™

Detailed Analysis
Project Zero

Report Date   6-21-91
Data Date   6-21-91
Program Manager Jim Lewis

Project Start   6-17-91
Project Finish
Page 1

------Task Description------ -Task ID--	--WBS--	Pla Sta Rev Sta	Pla Fin Rev Fin	BCWS	Cum to Date BCWP	ACWP	CV SV	CPI SPI	CVI SVI	BAC EAC	%A %E
Activity A		6-17-91 6-17-91	6-26-91 6-26-91	1200	768	960	-192 -432	0.80 0.64	-0.25 -0.36	1920 2160	50 40
Activity B		6-17-91 6-17-91	7-08-91 7-08-91	1680	1673	1680	-7 -7	0.99 0.99	0.00 0.00	4080 4080	41 41
Activity C		6-27-91 6-27-91	7-01-91 7-01-91				0 0	0.00 0.00	0.00 0.00	480 480	0
Activity D		6-27-91 6-27-91	7-08-91 7-08-91				0 0	0.00 0.00	0.00 0.00	2000 2000	0
Activity E		7-09-91 7-09-91	7-15-91 7-15-91				0 0	0.00 0.00	0.00 0.00	1000 1000	0
Activity F		7-16-91 7-16-91	7-18-91 7-18-91				0 0	0.00 0.00	0.00 0.00	600 600	0
Activity G		7-09-91 7-09-91	7-22-91 7-22-91				0 0	0.00 0.00	0.00 0.00	6400 6400	0
Finish			7-22-91 7-22-91								
				2880	2441	2640	-199 -439			16480 16720	

Section Five

Key Factors for Success

Chapter 14

Balancing Strategy and Tactics in Project Implementation*

It is the rare project manager who is a brilliant strategist *and* a skilled tactician, but to manage projects successfully, both capabili-

* This chapter was originally published by *Sloan Management Review*, Fall 1987, and is reprinted here by permission from Dennis P. Slevin, University of Pittsburgh Jeffrey K. Pinto, University of Cincinnati.

The authors wish to acknowledge the comments of Robert W. Zmud and an anonymous reviewer on a draft of this article.

ties must be brought to bear. The authors propose 10 critical success factors for projects, break them down into strategic and tactical subgroups, and place that model in a project-life-cycle framework. They discuss what problems are likely to occur if a project is well formulated strategically but mishandled tactically, or well executed but poorly conceived. *Ed.*

There is many a slip 'twixt the cup and the lip.

Successful project implementation is complex and difficult. Project managers must pay attention simultaneously to a wide variety of human, financial, and technical factors—and they are often made responsible for project outcome without being given sufficient authority, money, or manpower.

Project-based work tends to be very different from other organizational activities. Projects usually have a specific goal or goals, a defined beginning and end, and a limited budget. Often developed by a team of individuals with special expertise, projects usually consist of a series of complex tasks requiring high levels of coordination.

Perhaps not surprisingly, the project manager's job is characterized by role overload, frenetic activity, and superficiality. He or she needs tools that will help to identify critical issues and to prioritize them over the life of the project.

Project management tools must acknowledge that the manager is of necessity a generalist as well as a specialist: he or she must know how to *plan* effectively and *act* efficiently. Unfortunately, the "dreamers" who are effective strategists often lack the operational skills to realize their plans. Likewise, project managers who are uncomfortable with planning prefer to address concrete, well-defined problems. Balancing the interplay between planning and action-strategy and tactics may be a project manager's most important job.

Despite the fact that many project managers are uneasy with either the strategic or the tactical side of their work, project management research to date has generally failed to address this important issue. This article provides some conceptual tools designed to do so. It proposes 10 project management "critical success factors," defines their relationship to one another, and describes how they fit into a strategic-tactical framework. In addition, it pinpoints errors likely to occur if strategy is well managed but tactics are not, and vice versa. Finally, it offers some pragmatic advice about strategic and tactical project management.

The Project Life Cycle

The concept of a *project life cycle* provides a useful framework for looking at project dynamics over time. The idea is familiar to most managers; it is used to conceptualize work stages and the budgetary and organizational resource requirements of each stage.[1] As Figure 14.1 shows, this frame of reference divides projects into four distinct phases of activity.

☞ **Conceptualization.** The initial project stage. Top managers determine that a project is necessary. Preliminary goals and alternative project approaches are specified, as are the possible ways to accomplish these goals.

☞ **Planning.** The establishment of formal plans to accomplish the project's goals. Activities include scheduling, budgeting, and allocation of other specific tasks and resources.

☞ **Execution.** The actual "work" of the project. Materials and resources are procured, the project is produced, and performance capabilities are verified.

☞ **Termination.** Final activities that must be performed once the project is completed. These include releasing resources, transferring the project to clients, and, if necessary, reassigning project team members to other duties.

As Figure 14.1 shows, the project life cycle is useful for project managers because it helps to define the level of effort needed to perform the tasks associated with each stage. During the early stages, requirements are minimal. They increase rapidly during late planning and execution and diminish during termination. Project life cycles are also helpful because they provide a method for tracking the status of a project in terms of its stage of development.

Figure 14.1. Stages in the Project Life Cycle.

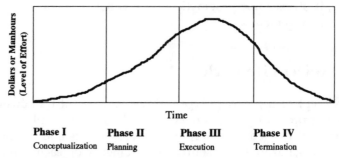

Adams and Barndt, "Behavioral Implications of the Project Life Cycle," in Project Management Handbook, ed. Cleland and King
Copyright (c) 1983 by Van Nostrand Reinhold Co., Inc.
Reprinted by permission of the publisher.

Project Critical Success Factors

In recent years the authors and other researchers have focused on identifying those factors most critical to project success and have generated both theoretical models and lists of "success" factors.[2] Through a recent study, we have developed and refined a set of critical success factors that we believe will make conceptual sense to managers, and that is general enough to be supported across a wide range of project types.[3] As we shall see, these factors fit into a broader framework that models the dynamic project implementation process. They have also led to the development of a Project Implementation Profile (PIP) that can be used to monitor and update the factors' status throughout a project's life. First, though, we should define the factors.[4]

☞ **Project Mission.** Initial clarity of goals and general direction.

☞ **Top Management Support.** Willingness of top management to provide the necessary resources and authority or power for project success.

☞ **Project Schedule/Plans.** Detailed specification of the individual action steps required for project implementation.

☞ **Client Consultation.** Communication and consultation with, and active listening to, all affected parties.

☞ **Personnel.** Recruitment, selection, and training of the necessary personnel for the project team.

☞ **Technical Tasks.** Availability of the required technology and expertise to accomplish the specific technical action steps.

☞ **Client Acceptance.** The act of "selling" the final project to its intend users.

☞ **Monitoring and Feedback.** Timely provision of comprehensive control information at each stage in the implementation process.

☞ **Communication.** Provision of an appropriate network and necessary data to all key actors in the project implementation.

☞ **Troubleshooting.** Ability to handle unexpected crises and deviations from plan.

A fifty-item instrument has been developed to measure a project's score on each of the ten factors in comparison to over 400 projects studied. The Project Implementation Profile provides a quantitative way of quickly profiling a project on these ten key factors.

As Figure 14.2 shows, we have developed a framework of project implementation based on the 10 factors. This framework is intended to demonstrate that these 10 factors are not only all critical to project success, but that there is also a relationship *among* the factors. In other words, these factors must be examined in relation to each other as well as to their individual impact on successful implementation. Conceptually, the factors are sequenced logically rather than randomly. For example, it is important to set goals or

Figure 14.2. Ten Key Factors of the Project Implementation Profile.

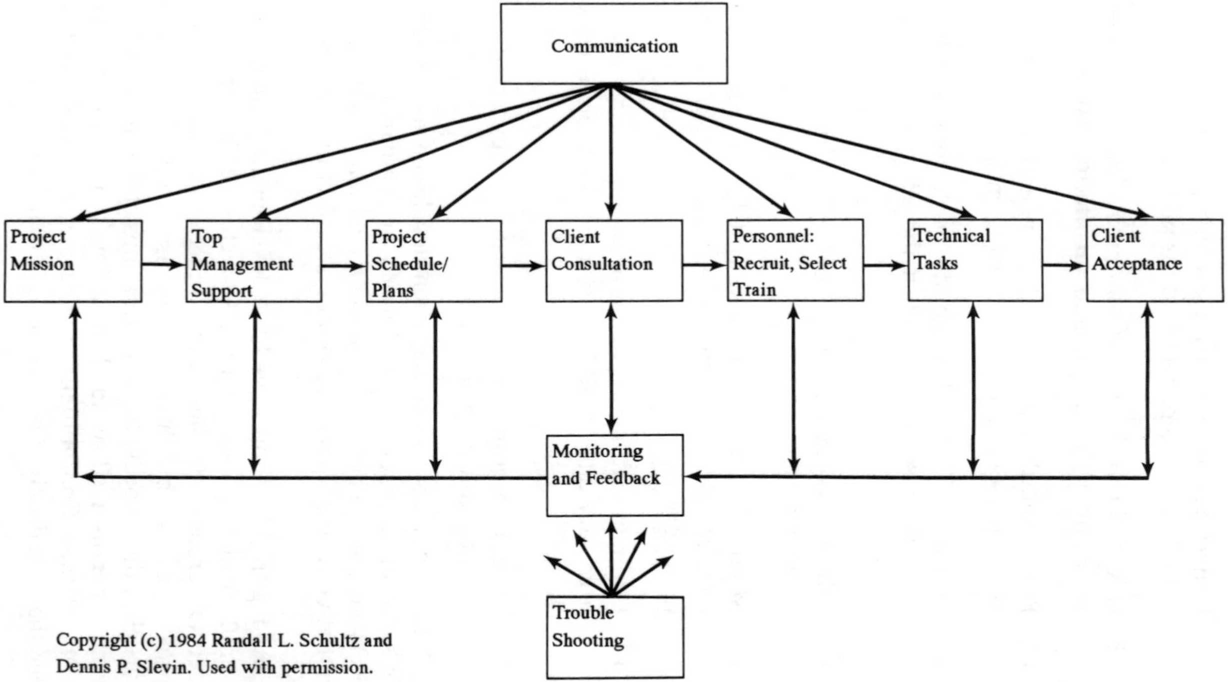

Communication

Client Acceptance

Technical Tasks

Personnel: Recruit, Select Train

Client Consultation

Monitoring and Feedback

Trouble Shooting

Project Schedule/ Plans

Top Management Support

Project Mission

define the mission and benefits of the program before seeking top management support. Similarly, unless consultation with clients occurs early in the process, chances of subsequent client acceptance will be lowered. In actual practice, considerable overlap can occur among the various factors, and their sequencing is not absolute. The arrows in the model represent information flows and sequences, not causal or correlational relationships.

As Figure 14.2 shows, in addition to the seven factors that can be laid out on a sequential critical path, three additional factors are hypothesized to play a more overriding role in the project implementation. These factors, monitoring and feedback, communication, and trouble shooting, must all necessarily be present at each point in the implementation process. Further, a good argument could be made that these three factors are essentially different facets of the same general concern (i.e., project communication). Communication is vital for project control, for problem solving, and for maintaining beneficial contacts with both clients and the rest of the organization.

Strategy and Tactics

As one moves through the 10-factor model, it becomes clear that the factors' general characteristics change. The first three (mission, top management support, and schedule) are related to the early, "planning" phase of project implementation. The other seven are concerned with the actual implementation or "action" of the project. These planning and action elements can usefully be considered *strategic*—the process of establishing overall goals and of planning how to achieve those goals-and *tactical*—using human, technical, and financial resources to achieve strategic ends. Briefly, the critical success factors of project implementation fit into a strategic/tactical breakout in the following way:

☞ **Strategic:** mission, top management support, project schedule/plans.

☞ **Tactical:** client consultation, personnel, technical tasks, client acceptance, monitoring and feedback, communication, trouble shooting.

Strategy and Tactics over Time

While both strategy and tactics are essential for successful project implementation, their importance shifts as the project moves through its life cycle. Strategic issues are most important at the beginning, tactical issues gain in importance toward the end. There should, of course, be continuous interaction and testing between the two-strategy often changes in a dynamic corporation, so regular monitoring is essential. Nevertheless, a successful project manager must be able to make the transition between strategic and tactical considerations as the project moves forward.

As Figure 14.3 shows, a recent study of more than 400 projects charted the shifting balance between strategic and tactical issues over the project's life cycle.[5] The "importance" value was measured by regression beta weights showing the relationships among strategy, tactics, and project success over the life cycle stages. During the two early stages, conceptualization and planning, strategy is significantly more important to project success than tactics. As the project moves toward the final stage, they achieve almost equal importance. Throughout the project, initial strategies and goals continue to "drive" or shape tactics.

These changes have important implications. A project manager who is a brilliant strategist but an ineffective tactician has a strong potential for committing certain types of errors as the project moves forward. These errors may occur after substantial resources have been expended. In contrast, the project manager who is excellent at tactical execution but weak in strategic thinking has a potential for committing different kinds of errors. These will more likely occur early in the process, but may remain undiscovered because of the manager's effective execution.

Figure 14.3. Changes in Strategy and Tactics across the Project Life Cycle (n = 418).

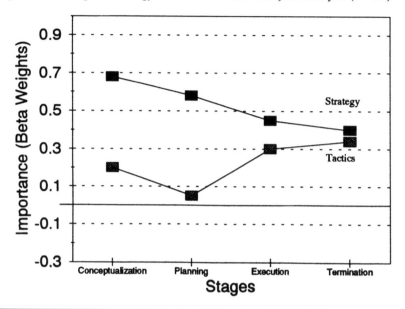

Strategic and Tactical Performance

Figure 14.4 shows the four possible combinations of strategic and tactical performance and the kinds of problems likely to occur in each scenario. The values "high" and "low" represent strategic and tactical *quality*, i.e., effectiveness of operations performed.

A *Type I* error occurs when an action that should have been taken was not. Consider a situation in which strategic actions are adequate and suggest development and implementation of a project. A Type I error has occurred if tactical activities are inadequate, little action is subsequently taken, and the project is not developed.

A *Type II* error happens if an action is taken when it should not have been. In practical terms, a Type II error is likely to occur if the project strategy is ineffective or inaccurate, but goals and schedules are implemented during the tactical stage of the project anyway.

Figure 14.4. Strategy/Tactics Effectiveness Matrix.

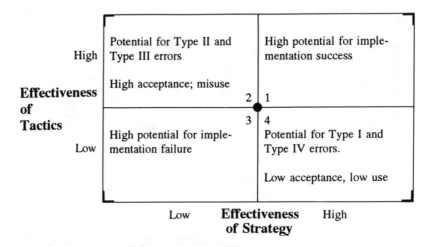

Type I error: Not taking an action when one should be taken.
Type II error: Taking an action when none should be taken.
Type III error: Taking the wrong action (solving the wrong problem).
Type IV error: Addressing the right problem, but solution is not used.

Source: Schultz, Slevin and Pinto (1987).[8]

A *Type III* error can be defined as solving the wrong problem, or "effectively" taking the wrong action. In this scenario, a problem is identified, or a project is desired, but because of a badly performed strategic sequence, the wrong problem is isolated, so the implemented project has little value-it does not address the intended target. Such situations often involve large expenditures of human and budgetary resources (tactics) for which there is inadequate initial planning and problem recognition (strategy).

Type IV is the final kind of error common to project implementation: the action taken does solve the right problem, but the solution is not used. That is, if project management correctly identifies a problem, proposes an effective solution, and implements that solution using appropriate tactics—but the project is not used by the client for whom it was intended—then a Type IV error has occurred.

As Figure 14.4 suggests, each of these errors is most likely to occur given a particular set of circumstances.

☞ **Cell 1. High Strategy/High Tactics.** Cell 1 is the setting for projects rated effective in carrying out both strategy and tactics. Not surprisingly, most projects in this situation are successful.

☞ **Cell 3. Low Strategy/Low Tactics.** The reciprocal of the first is the third cell, where both strategic and tactical functions are inadequately performed. Projects in this cell have a high likelihood of failure.

☞ **Cell 4. High Strategy/Low Tactics.** The results of projects in the first two cells are intuitively obvious. Perhaps a more intriguing question concerns the likely outcome for projects found in the "off diagonal" of Figure 14.4, namely, High Strategy/Low Tactics and Low Strategy/ High Tactics.

In Cell 4, the project strategy is effective but subsequent tactics are ineffective. We would expect projects in this cell to have a strong tendency toward "errors of inaction" such as low acceptance and low use by organization members or clients for whom the project was intended. Once a suitable strategy has been determined, little is done in the way of tactical follow-up to operationalize the goals of the project or to "sell" the project to its prospective clients.

☞ **Cell 2. Low Strategy/High Tactics.** The final cell reverses the preceding one. Here, project strategy is poorly conceived or planning is inadequate, but tactical implementation is well managed. Projects in this cell often suffer from "errors of action." Because of poor strategy, a project may be pushed into implementation even though its purpose has not been clearly defined. In fact, the project may not even be needed. However, tactical follow-up is so good that the inadequate or unnecessary project is implemented.

The managerial attitude is to "go ahead and do it"; not enough time is spent early in the project's life assessing whether the project is needed and developing the strategy.

Case Study Illustrations

In the section that follows, we discuss four instances in which strategic and tactical effectiveness were measured by project participants using the Project Implementation Profile. We caution that the results were reported in three instances by only one observer—the project manager—so they are obviously not meant as evidence in support of an argument, but rather as an illustration of distinct project-outcome types. In each case, a ten-factor profile is provided, using the actual scores from the PIP based on input from the project managers.

High Strategy/High Tactics: The New Alloy Development

One department of a large organization was responsible for coordinating the development and production of new stainless steel alloys for the automotive exhaust market. This task meant overseeing the effort of the metallurgy, research, and operations departments. The project grew out of exhaust component manufacturers' demands for more formable alloys. Because this product line represented a potentially significant portion of the company's market, the project was given high priority.

As Figure 14.5A demonstrates, the scores for this project as assessed by the project team member were uniformly high across the ten critical success factors. Because of the importance of the project, its high priority was communicated to all personnel, and this led to a strong sense of project mission and top management support. The strategy was clear and was conveyed to all concerned parties, including the project team, which was actively involved in early planning meetings. Because the project team would include personnel from research, metallurgy, operations, production, and commercial departments, great care was taken in its selection and coordination. Use throughout the project team of action plans and daily exception

Figures 14.5A - 14.5D

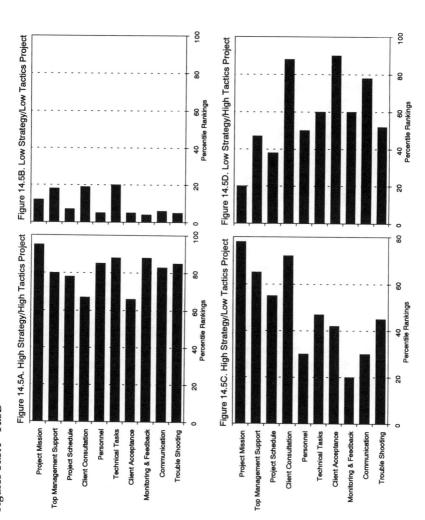

Figure 14.5A. High Strategy/High Tactics Project

Figure 14.5B. Low Strategy/Low Tactics Project

Figure 14.5C. High Strategy/Low Tactics Project

Figure 14.5D. Low Strategy/High Tactics Project

reports was reflected in high scores on Technical Tasks and Trouble Shooting.

In the new alloy development project, a strong, well-conceived strategy was combined with highly competent tactical follow-up. The seeds of project success were planted during the conceptual and planning stages and were allowed to grow to their potential through rigorous project execution. Success in this project can be measured in terms of technical excellence and client use, as well as project team satisfaction and commercial profitability. In a recent follow-up interview, a member of a major competitor admitted that the project was so successful that the company still has a virtual lock on the automotive exhaust market.

Low Strategy/Low Tactics: The Automated Office

A small, privately owned company was attempting to move from a no automated, paper system to a fully-integrated, automated office that would include purchasing, material control, sales order, and accounting systems. The owner's son, who had no previous experience with computers, was hired as MIS director. His duties consisted of selecting hardware and software, directing installation, and learning enough about the company to protect the family's interests. Figure 14.5B shows a breakdown of he 10 critical success factors as viewed by a project team member.

Several problems emerged immediately. Inadequate "buy-in" on the part of organization members, perceived nepotism, and lack of interaction with other top managers in purchasing decisions were seen as problems while the project was still in its strategy phase. A total lack of a formal schedule or implementation plan emphasized other strategic inadequacies destined to lead to tactical problems as well.

Tactically, the project was handled no better. Other departments that were expected to use the system were not consulted about their specific needs; the system was simply forced upon them. Little effort was made to develop project control and trouble-shooting mechanisms, perhaps as a direct result of inadequate scheduling.

Project results were easy to predict. As the team member indicated and Figure 5B reinforces, the project was over budget, behind schedule, and coolly received—all in all, an expensive failure. The owner's son left the company, the manager of the computer department was demoted, the mainframe computers were found to be wholly inadequate and were sold, and upper management forfeited a considerable amount of employee goodwill.

High Strategy/Low Tactics: The New Bank Loan Setup

The purpose of this project was to restructure the loan procedures used at a major bank. The project was intended to eliminate duplicate work done by branches and the servicing department and to streamline loan processes. These goals were developed and strongly supported by upper management, which had clearly conveyed them to all concerned parties. The project was kicked off with a great deal of fanfare; there was a high expectation of speedy and successful completion. Trouble started when the project was turned over to a small team that had not been privy to the initial planning, goal setting, and scheduling meetings. In fact, the project team leader was handed the project after only three months with the company.

Project tactics were inadequate from the beginning. The team was set up without any formal feedback channels and with few communication links with either the rest of the organization or top management. The project was staffed on an ad hoc basis, often with nonessential personnel from other departments. This staffing method resulted in a diverse team with conflicting loyalties. The project leader was never able to pull the team together.

As the project leader put it, "Although this project hasn't totally failed, it is in deep trouble." Figure 14.5C illustrates the breakdowns for the project as reported by two team members. Almost from the start of its tactical phase, the project suffered from the team's inability to operationalize the initial goals. This failure caused frustration both within the project team and throughout the rest of the organization. The frustration resulted from having a clear idea of the initial goals without having prescribed the means to achieve them. As of this writing, the project continues to stagger along, with cost overruns and constantly revised schedules. Whether or not it achieves its final performance goals, this project will be remembered with little affection by the rest of the organization.

Low Strategy/High Tactics: The New Appliance Development

A large manufacturing company initiated the development of a new kitchen appliance to satisfy what upper management felt would be a consumer need in the near future. The project was perceived as the pet idea of a divisional president and was rushed along without adequate market research or technical input from the R&D department. A project team was formed to develop the product and rush it to the marketplace.

Figure 5D shows the breakdowns of the 10 critical success factors for this project. Organizational and project team commitment was low. Other members of upper management felt the project was being pushed along too fast and refused to get behind it. Initial planning and scheduling developed by the divisional president and his staff were totally unrealistic.

What happened next was interesting. It was turned over to an experienced, capable manager who succeeded in taking the project, which had gotten off to such a shaky start, and successfully implementing it. He reopened channels of communication within the organization, bringing R&D and marketing on board. He met his revised schedule and budget, using trouble-shooting and control mechanisms. Finally, he succeeded in getting the project to the market in a reasonable time frame.

In spite of the project manager's effective tactics, the product did not do well in the market. As it turned out, there was little need for it at the time, and second-generation technology would make it obsolete within a year. This project was highly frustrating to project team members, who felt, quite correctly, that they had done everything possible to achieve success. Through no fault of their own, this project was doomed by the poor strategic planning. All the tactical competence in the world could not offset the fact that the project was poorly conceived and indifferently supported, resulting in an "error of action."[6]

Implications for Managers

These cases, and the strategy/tactics effectiveness matrix, suggest practical implications for managers wishing to better control project implementation.

☞ *Use a multiple-factor model.* Project management is a complex task requiring attention to many variables. The more specific a manager can be regarding the definition and monitoring of those variables, the greater the likelihood of successful project outcome. It is important to use a multiple-factor model to do this, first to understand the variety of factors affecting project success, then to be aware of their relative importance across project implementation stages.[7] This article offers such a model: 10 critical success factors that fit into a process framework of project implementation; within the framework, different factors become more critical to project success at different points in the project life cycle.

Additionally, both the project team and clients need to perform regular assessments to determine the "health" of the project. The time for accurate feedback is when the project is beginning to develop difficulties that can be corrected, not down the road when the troubles have become insurmountable. Getting the project team as well as the clients to perform status checks has the benefit of giving insights from a variety of viewpoints, not just that of the project manager. Further, it reinforces the goals the clients have in mind, as well as their perceptions of whether the project satisfies their expectations.

☞ *Think strategically early in the project life cycle.* It is important to consider strategic factors early in the project life cycle, during conceptualization and planning. As a practical suggestion, organizations implementing a project should bring the manager and his or her team on board early. Many managers make the mistake of not involving team members in early planning and conceptual meetings, perhaps assuming that the team members should only concern themselves with their specific jobs. In fact, it is very important that at an early stage both the manager *and* the team members "buy in" to the goals of the project and the means to achieve those goals. The more team members are aware of the goals, the greater the likelihood of their taking an active part in monitoring and trouble shooting.

☞ *Think more tactically as the project moves forward in time.* As Figure 14.4 shows, in the later project stages, strategy and tactics are of almost equal importance to project implementation success. Consequently, it is important that the project manager shift the team's emphasis from "What do we do?" to "How do we want to do it?" The specific critical success factors associated with project tactics tend to reemphasize the importance of focusing on the "how" instead of the "what." Factors such as personnel, communication, and monitoring are concerned with better managing specific action steps in the project implementation process. While we argue that it is important to bring the project team on board during the initial strategy phase, it is equally important to manage their shift into a tactical, action mode in which their specific project duties are performed to help move the project toward completion.

☞ *Consciously plan for and communicate the transition from strategy to tactics.* Project monitoring will include an open, thorough assessment of progress at several stages of implementation. The assessment must acknowledge that the transition from a strategic to a tactical focus introduces an additional set of critical success factors.

Project managers should regularly communicate with team members about the shifting status or focus of the project. Communication reemphasizes the importance of a joint effort, and it reinforces the status or the project relative to its life cycle. The team is kept aware of the degree of strategic versus tactical activity necessary to move the project to the next life-cycle stage. Finally, communication helps the manager to track the various activities performed by the project team, making it easier to verify that strategic vision is not lost in the later phases of tactical operationalization.

☞ *Make strategy and tactics work for you and your project team.* Neither strong strategy nor strong tactics by themselves will ensure project success. When strategy is strong and tactics are weak, there is a potential for creating projects that never get off the ground. Cost and schedule overruns, along with general frustration, are often the side effects of projects that encounter "errors of inaction." On the other hand, a project that starts off with a weak or poorly conceived strategy and receives strong subsequent tactical operationalization is likely to be successfully implemented, but to address the wrong problem. New York advertising agencies can tell horror stories of advertising campaigns that were poorly conceived but still implemented, sometimes costing millions of dollars, and that were ultimately judged disastrous and scrubbed.

In addition to having project strategy and tactics working together, it is important to remember (again following Figure 14.3) that strategy should be used to "drive" tactics. Strategy and tactics are not independent of each other. At no point do strategic factors become unimportant to project success; instead, they must be continually assessed and reassessed over the life of the project in light of new project developments and changes in the external environment.[8]

Endnotes

[1] The four-stage project live cycle is based on work by J. Adams and S. Barndt, "Behavioral Implications of the Project Life Cycle," in *Project Management Handbook,* eds., D. I. Cleland and W. R. King (New York: Van Nostrand Reinhold, 1983), pp. 222–244.

[2] For an alternative methodology for the development of critical success factors for the implementation of organizational systems, see the work of M. Shank, A. Boynton, and R. W. Zmud, "Critical-Success Factor Analysis as a Methodology for MIS Planning," *MIS Quarterly*, June 1985, pp. 121–129. See further, A. Boynton and R. W. Zmud. "An Assessment of Critical Success Factors," *Sloan Management Review*, Summer 1984, pp. 17–27.

[3] D. P. Slevin and J. K. Pinto, "The Project Implementation Profile: New Tool for Project Managers," *Project Management Journal*, 17 (1986): 57–70.

[4] J. K. Pinto and D. P. Slevin, "Critical Factors in Successful Project Implementation," *IEEE Transactions on Engineering Management*, EM–34, February 1987, pp. 22–27.

[5] J. K. Pinto, "Project Implementation: A Determination of Its Critical Success Factors, Moderators, and Their Relative Importance Across Stages in the Project Life Cycle" (Pittsburgh, PA: University of Pittsburgh, unpublished doctoral dissertation, 1986).

[6] Pinto (1986).

[7] For a copy of the full 100-Item Project Implementation Profile, see Slevin and Pinto (1986).

[8] R. L. Schultz, D. P. Slevin, and J. K. Pinto, "Strategy and Tactics in a Process Model of Project Implementation," *Interfaces*, May–June 19887, pp. 34–46.

Chapter 15

The Project Implementation Profile[*]

Introduction

The project implementation profile (PIP) was developed by Slevin and Pinto to assist project managers in applying their model of balancing strategy and tactics as outlined in Chapter 14. So that read-

---

[*]   *The Project Implementation Profile* is copyright 1992 by Xicom, Inc., RR #2, Woods Road, Tuxedo, NY 10987. Reprint permission granted.

Copies of the profile can be ordered from the above address. Tel. 914-351-4752; 800-759-4266; FAX: 914-351-4762.

---

ers of this handbook will know how the profile works, the questions contained in the PIP follow. Each factor is measured by circling the point on a 7-point scale at which the item falls, then totalling the items for each factor to get an overall score. This score is then compared to a table of norms, to allow the project manager to see how his or her project compares with others.

The norms are not contained in this chapter. They are only available as part of the complete PIP instrument, which can be ordered from XICOM.

The numerical scale on which each item is rated is a 7-point scale as shown below:

Strongly Disagree			**Neutral**			**Strongly Agree**
1	2	3	4	5	6	7

Factor 1

Project Mission

Initial clarity of goals and general direction.

1. The goals of the project are in line with the general goals of the organization.
2. The basic goals of the project were made clear to the project team.
3. The results of the project will benefit the parent organization.
4. I am enthusiastic about the chances for success of this project.
5. I am aware of and can identify the beneficial consequences to the organization of the success of this project.

Factor 2

Top-Management Support

Willingness of top management to provide the necessary resources and authority/power for project success.

1. Upper management will be responsive to our requests for additional resources, if the need arises.

2. Upper management shares responsibility with the project team for ensuring the project's success.

3. I agree with upper management on the degree of my authority and responsibility for the project.

4. Upper management will support me in a crisis.

5. Upper management has granted us the necessary authority and will support our decisions concerning the project.

Factor 3

Project Schedule/Plan

A detailed specification of the individual action steps required for project implementation.

1. We know which activities contain slack time or slack resources that can be utilized in other areas during emergencies.

2. There is a detailed plan (including time schedules, milestones, manpower requirements, etc.) for the completion of the project.

3. There is a detailed budget for the project.

4. Key personnel needs (who, when) are specified in the project plan.

5. There are contingency plans in case the project is off schedule or off budget.

Factor 4

Client Consultation

Communication, consultation, and active listening to all impacted parties.

1. The clients were given the opportunity to provide input early in the project development stage.

2. The clients (intended users) are kept informed of the project's progress.

3. The value of the project has been discussed with the eventual clients.

4. The limitations of the project have been discussed with the clients (what the project is *not* designed to do).

5. The clients were told whether or not their input was assimilated into the project plan.

Factor 5

Personnel

Recruitment, selection, and training of the necessary personnel for the project team.

1. Project team personnel understand their role on the project team.

2. There is sufficient manpower to complete the project.

3. The personnel on the project team understand how their performance will be evaluated.

4. Job descriptions for team members have been written and distributed and are understood.

5. Adequate technical and/or managerial training (and time for training) are available for members of the project team.

Factor 6

Technical Tasks

Availability of the required technology and expertise to accomplish the specific technical action steps.

1. Specific project tasks are well managed.

2. The project engineers and other technical people are competent.

3. The technology that is being used to support the project works well.

4. The appropriate technology (equipment, training programs, etc.) has been selected for project success.

5. The people implementing this project understand it.

Factor 7

Client Acceptance

The act of "selling" the final project to its ultimate intended users.

1. There is adequate documentation of the project to permit easy use by the clients (instructions, etc.).

2. Potential clients have been contacted about the usefulness of the project.

3. An adequate presentation of the project has been developed for clients.

4. Clients know whom to contact when problems or questions arise.

5. Adequate advance preparation has been done to determine how best to "sell" the project to clients.

Factor 8

Monitoring and Feedback

Timely provision of comprehensive control information at each stage in the implementation process.

1. All important aspects of the project are monitored, including measures that will provide a complete picture of the project's progress (adherence to budget and schedule, manpower and equipment utilization, team morale, etc.).

2. Regular meetings to monitor project progress and improve the feedback to the project team are conducted.

3. Actual progress is regularly compared with the project schedule.

4. The results of project reviews are regularly shared with all project personnel who have impact upon budget and schedule.

5. When the budget or schedule requires revision, input is solicited from the project team.

Factor 9

Communication

The provision of an appropriate network and necessary data to all key actors in the project implementation.

1. The results (decisions made, information received and needed, etc.) of planning meetings are published and distributed to applicable personnel.

2. Individuals/groups supplying input have received feedback on the acceptance or rejection of their input.

3. When the budget or schedule is revised, the changes *and* the reasons for the changes are communicated to all members of the project team.

4. The reasons for the changes to existing policies/procedures are explained to members of the project team, other groups affected by the changes, and upper management.

5. All groups affected by the project know how to make problems known to the project team.

Factor 10

Troubleshooting

Ability to handle unexpected crises and deviations from the plan.

1. The project leader is not hesitant to enlist the aid of personnel not involved in the project in the event of problems.

2. Brainstorming sessions are held to determine where problems are most likely to occur.

3. In case of project difficulties, project team members know exactly where to go for assistance.

4. I am confident that problems that arise can be solved completely.

5. Immediate action is taken when problems come to the project team's attention.

Project Performance

In addition to the previous 10 factors, please give your assessment of overall project performance by responding to the following 12 items. Total your overall score for all 12 items at the bottom of the following page.

1. This project has/will come in on schedule.

2. This project has/will come in on budget.

3. The project that has been developed works (or, if still being developed, looks as if it will work).

4. The project will be/is used by its intended clients.

5. This project has directly benefitted/will directly benefit the intended user through either increasing efficiency or employee effectiveness.

6. Given the problem for which it was developed, this project seems to do the best job of solving that problem—i.e., it was the best choice among the set of alternatives.

7. Important clients directly affected by this project will make use of it.

8. I am/was satisfied with the process by which this project is being/was completed.

9. We are confident that nontechnical start-up problems will be minimal, because the project will be readily accepted by its intended users.

10. Use of this project has led/will lead directly to improved or more effective decision-making or performance for the clients.

11. This project will have a positive impact on those who make use of it.

12. The results of this project represent a definite improvement in performance over the way clients used to perform these activities.

Chapter 16

Factors in Project Success[*]

Introduction

Middle managers today are expected to possess project management skills and utilize them in conjunction with total quality management and continuous quality improvement. As director of a middle management continuing education program at a major university, the author added a project management seminar to the

---

[*] This chapter was originally titled "Project Management Factors and Performance as Perceived by Middle Managers" by David Antonioni, Ph.D., Director of The Middle Management Program Series, School of Business-Management Institute, University of Wisconsin-Madison.

middle management program series. As part of the seminar, managers assessed their project management process using the Project Implementation Profile (PIP) questionnaire developed by Slevin and Pinto (1986). The project profile was validated primarily with project managers, project team leaders, and technical and administrative project members. Slevin and Pinto developed 10 reliable and valid project management factors for project implementation, which were later found to be significantly correlated with project performance (Pinto and Slevin, 1988). The author decided to use the profile with middle managers to determine if Slevin and Pinto's findings could be replicated. In addition, the author wanted to investigate the relationship between project performance and three additional factors: organizational climate, supervision, and peer relations. What follows is a brief discussion of the project management profile; reasons for investigating organizational climate, supervision, and peer relations; and the methods used to gather data for this study. Results and implications for middle managers using project management are also presented.

Slevin and Pinto (1986) established 10 project management factors: (1) project mission—clarity of initial goals and general directions, (2) top management support—willingness of top management to provide resources, authority and power for project success, (3) project schedule/plan—detailed individual action plans for project implementation, (4) client consultation—communication and active listening to all client stakeholders, (5) personnel—recruitment, selection, and training for project members, (6) technical tasks—availability of technology and expertise required to do the job, (7) client acceptance—selling the final project to end users, (8) monitoring and feedback—timely information to help control the project at each stage, (9) communication—establishment of useful network and necessary data for all project stakeholders, and (10) troubleshooting—capability to handle unexpected crises and deviations from the plan.

In a subsequent study, Slevin and Pinto (1987) discussed four stages of the project life cycle: conceptualization, planning, execution and termination. They claimed that project mission was the most important factor in all four phases of project management because the success of a project was critically dependent on the clarity of the project goals. Client consultation was the second most important factor because clients' needs and expectations were strongly linked to project goals (Pinto and Slevin, 1989).

Project mission and client consultation should be the most important predictors of project performance. It is hard to imagine that projects would be successful without clear goals, general directions, or active consultation with the project client. Would middle managers perceive these factors as most important? The purpose of this study was to determine which project management factors middle managers associated with project performance.

Pinto and Slevin (1988) discovered that the 10 project management factors mentioned above correlated positively with project success. They defined project success with the following measures: (1) Did the project attain its goals? (2) Was it done on time? (3) Was it done within budget? and (4) Was the client satisfied? These are valid measures. However, there were additional factors that the author believed should be investigated. Therefore, this study investigated the relationships that organizational climate, supervision, and peer relations have with project performance.

The project management process exists within the context of an organizational climate and should therefore have a significant relationship with project performance. Webster defines climate as "the environmental conditions characterizing a group." Conditions in the work environment should have a strong relationship with project performance, thus, middle managers working in organizations that have a poor organizational climate should indicate less project success than managers working in better organizational climates. It seems unlikely that individuals would rate project success high when responses indicate a poor organizational climate. There is a lack of opportunity for input in decision making, information is not shared, abilities and skills are not utilized, the organization is not quick to use improved work methods, or different work units do not plan together and coordinate their efforts.

Project performance could also depend on the quality of the work relationship with one's supervisor. If middle managers are highly satisfied with their supervisors, to what extent does this affect project performance? Supervisors who demonstrate support, encourage teamwork, maintain high performance standards, and provide training should affect the middle manager's project performance. This study investigated the strength of the relationship between supervision and project performance.

Finally, middle managers working on projects also interact with their work-unit peers. Depending on the project, work-unit peers may or may not be on a project team. Work-unit peer relation factors such as approachability, encouraging teamwork, maintaining high performance standards, and finding work improvement methods could affect project performance. This study also examined the nature of the relationship between work-unit peer relations and project performance.

Methods

Participants: 128 middle managers from Midwest companies participated in this study. The average amount of managerial experience was 7.6 years. Thirty-two percent of the managers were female and 68 percent were male. Sixty-five percent of the participants were from manufacturing companies and 35 percent were from service companies. Examples of position titles included: Manufacturing Engineer Manager, Manager of Quality and Technical Service, Manager of System Development, Manager of Marketing Services, and Corporate Distribution Manager.

Measures: Slevin and Pinto's 50-item Project Implementation Profile (PIP) questionnaire was administered to all participants. Ten PIP factors were used to assess the project management process. Each factor had five Likert-scaled questions.

A 12-item questionnaire was used to measure project success (Pinto and Slevin, 1988). These items asked managers to assess the extent to which projects were on schedule, within budget, and satisfied customers' performance expectations.

A 26-item Quick Assessment Survey (QAS) measured organizational climate, supervision and peer relations. The QAS items were taken from Likert's Survey of Organizations (Denison, 1984). Climate included the following: Receptivity to suggestions and decision making; utilization of one's skills and abilities; job expectations; recognition and respect from one's job; quick organizational response to improve work methods, clarity of objectives and goals; sensible organization of work activities and different work units planning and coordinating their work efforts. Supervision included items on support, team building, goal setting, and training. Finally, peer relations assessed how much people are encouraged to work in a team, the team's performance standards, and team facilitation of finding ways to improve performance.

Procedure: Participants who attended a three-day seminar on project management were asked to complete the PIP and QAS as part of the seminar experience. All survey data was collected from seminars conducted during the 1991-92 academic calendar. Participants received summary results from both questionnaires and discussed them in the seminars.

Data Analysis: Correlation coefficient and coefficient alphas were computed for all factors. All factors possessed acceptable iternal consistency requirements. Correlations were computed beween dimensions project management, organizational environment, and project performance or success. Multiple regression was also conducted to determine the extent to which project factors, organizational climate, supervision, and peer relations predict project performance.

Results

The 10 PIP factors' coefficient alphas were as follows: project mission (.89), top management support (.88), project scheduling and planning (.82), client consultation (.91), personnel (.73), technical tasks (.80), client acceptance (.80), monitoring and feedback (.78), communication (.88), and trouble-shooting (.75). The project performance questionnaire had a coefficient alpha of .84. The QAS factors had the following coefficient alphas: climate (.86), supervision (.86), and peer relations (.85).

Monitoring feedback, project mission, and communication were rated as more positively correlated with project performance than the remaining factors. The relationship between client consultation and project performance was the weakest of all 10 project factors. Table 16.1 shows the correlations between project management factors and project performance.

Organizational climate, supervision, and peer relations were all significantly correlated with project performance. The strength of the relationship between organizational climate and project performance was one of the stronger relationships. Table 16.2 shows the correlation between organizational climate, supervision, and peer relations with project performance.

Table 16.1. Correlations between Project Management Factors and Project Performance.

Variable	Project Performance
Monitoring Feedback	.5210***
Project Mission	.4836***
Communication	.4469***
Troubleshooting	.4659***
Client Acceptance	.4101***
Technical Tasks	.3898***
Project Schedule	.3735***
Personnel	.3593***
Management Support	.3735***
Client Consultation	.2346*

*** <.001,   ** <.01,   * <.05   N = 116

Table 16.2. Correlations between Organizational Climate, Supervision, and Peer Relations with Project Success.

Variable	Project Performance
Organizational Climate	.4038***
Supervision	.1965*
Peer Relations	.2936**

*** <.001,   ** <.01, *   <.05   N = 128

Results of a multiple regression analysis with the 10 project management factors as a predictor of project performance indicate an adjusted R square of .35055 with an F value of 7.20735 (significant F < .001). This means that the 10 factors are capable of explaining 35 percent of the variation in project success. Pinto and Slevin (1989) employed a stepwise regression analysis and found that they could predict about 60 percent of the variance in project performance using the project factors. Table 16.3 shows the multiple regression standardized beta, T values, and significance for each of the ten project implementation factors. The T values indicate how reliable each individual variable is as a predictor of project performance. Larger values of T (positive or negative) mean that a given variable is a good predictor of project performance.

Table 16.3. Multiple Regression Analysis of Project Management Factors.

Variable	Standardized Beta	T	Sig T
Troubleshooting	.232550	2.205	.0296
Project mission	.261171	2.730	.0074
Project scheduling	.112348	1.056	.2934
Client consultation	−.150345	−1.444	.1518
Management support	.075182	.840	.4031
Technical task	.053234	.549	.5839
Communication	.073080	.638	.5247
Personnel	−.146619	−1.301	.1960
Client acceptance	.120477	1.015	.3123
Monitoring feedback	.172406	1.355	.1783

When organizational climate, supervision, and peer relations are added to the 10 project implementation factors, the adjusted R square is .43885, with F = 7.737 (significance of F < .001). Table 16.4 shows the results of the multiple regression, which included climate, supervision, and peer relations. This shows that adding organizational climate, supervision, and peer relations as predictors along with the 10 PIP factors results in a better prediction of project performance than using the 10 PIP factors alone. A regression analysis using organizational climate alone produced an adjusted R square of .11578 with F = 17.497 (significance of F < .001). This result indicates that organizational climate by itself is also a good predictor of project performance.

Table 16.4. Multiple Regression of Project Management Factors Plus Organizational Climate, Supervision, and Peer Relations

Variable	Standardized Beta	T	Sig T
Peer relations	.166731	1.828	.0705
Communication	.176638	1.659	.1004
Management support	.054948	.559	.5772
Project scheduling	.107465	1.078	.2837
Supervision	−.163582	-1.843	.0684
Project mission	.310789	3.497	.0007
Client consultation	−.131136	−1.347	.1811
Technical task	.029542	.324	.7468
Troubleshooting	.095127	.896	.3726
Personnel	-.248211	-2.342	.0212
Client acceptance	.095109	.851	.3970
Climate	.167173	1.494.	.1383
Monitoring feedback	.243577	2.057	.0423

Discussion

This study confirms Slevin and Pinto's previous findings that the 10 project implementation factors are significantly related to project performance. More important, it shows how middle managers in mid-size companies perceive relationships between project performance and project management factors. Middle managers concur with project managers that project mission is one of the most significantly positive factors associated with project performance. However, middle managers differ from project managers in that they do not see client consultation as positively correlated with project performance. Conceptually, project mission and client consultation should be the foundation for project performance success. The difference in managers' perceptions may be attributable to their roles. Middle managers have to be more concerned with monitoring and feedback when it comes to performance. Their roles require that they use appropriate network communication in the implementation of a project.

Organizational climate is an important factor in project success. Results of this study suggest that measures of organizational climate may be important to middle managers who are expected to conduct some of their work through a project management format. A middle manager's perceptions of his or her organization's decision-making process are important in light of continuous improvement projects. The extent to which an organization is quick to utilize improved work methods is equally important, especially with quality-improvement projects. The extent to which different work units plan and coordinate their work efforts will affect project performance. Based on the findings of this study it is recommended that organizational climate be assessed in conjunction with the Project Implementation Profile.

Future studies should further investigate how the middle managers spend their time developing strategies and tactics for the project implementation process. Slevin and Pinto (1987) stated that strategy involves examining the big picture and planning ways to ensure goal attainment. Tactics involve the day-to-day decisions on how to use immediate resources to achieve strategic goals. They suggest that project managers should use strategy in regard to project mission, client consultation, and project scheduling.

Middle managers should learn to insist on a clearly specified on-target project mission and consult with key clients at the beginning of a project. Middle managers may need to get more involved in the conceptualization and planning process to make sure they use strategies to gain top-management support and project scheduling. Middle managers may place too much emphasis on tactics and not enough on strategies because they are accustomed to tactical firefighting. Middle managers may need to learn how to limit tactics and spend more time upfront strategically planning projects.

The results of this study suggest that middle managers see project mission, personnel, monitoring, and feedback as the most significant individual predictors of project success. However, personnel, monitoring, and feedback should be part of tactical planning. Middle managers may need to spend more time in the conceptual and planning phase of project management.

Project management training offers middle managers an opportunity to improve their overall planning skills and improve the implementation of continuous improvement projects. This study has demonstrated some differences between project managers and middle managers. More middle managers will be expected to manage projects in the near future. To facilitate this, it appears that middle managers need to improve their strategic planning skills. Finally, organizational climate is important and should always be included in project management implementation assessments.

References

Denison, D.R., "Bringing Corporate Culture to the Bottom Line," *Organizational Dynamics*, Reprint, 5-22 (1984).

Pinto, J.K., and D.P. Slevin, "Critical Success Factors Across The Project Life Cycle," *Project Management Journal*, 67-72 (1989).

Pinto, J.K., and D.P. Slevin, "Project Success: Definitions and Measurement Techniques," *Project Management Journal*, 19(1), 67-72 (1988).

Slevin, D.P., and J.K. Pinto, "Balancing Strategy and Tactics in Project Implementation," *Sloan Management Review*, 29(1), 33-41 (1987).

Slevin, D.P., and J.K. Pinto, "The Project Implementation Profile: New Tool for Project Managers," *Project Management Journal*, XVII(4), 57-70 (1986).

Chapter 17

Causes of Project Failure

The High Cost of Project Failure

It has been estimated that nearly half of the work done in some organizations is of a project nature. That means, clearly, that if many projects are failures, the organization as a whole is on its way to disaster.

With Slevin's *Project Implementation Profile,* project managers can monitor some of the factors known to contribute to success. But what about the causes of failure? If these are understood, then presumably some steps can be taken to avoid them. At least that is the intention of this chapter, to alert the project manager to the typical causes of failure with prevention as the objective.

Causes and Recommendations for Solutions

Failure to properly define the problem. This is commented on in an earlier chapter. As Juran says, a project is a problem scheduled for solution. If the problem is not well understood, then we may make the classic error of developing the right solution to the wrong problem. This can be avoided by attempting to understand the *real* reason for doing the job, then writing a problem statement to reflect that objective.

For example, a group is given the assignment to relocate an office to another part of the building. They may see their job as just moving furniture and partitions. But what is the real intention of moving the office? Perhaps it is to achieve better coordination between the people being moved and those near the new location. Is there an optimum layout that will make the move more effective? Only by understanding the real purpose of the move can this question be answered or addressed.

The high cost of project failure

Planning was based on insufficient data. As an example of this, an engineering project is planned that will involve significant use of a test facility. What the team doesn't know is that the test facility is to be relocated at exactly the time when their testing is scheduled to start.

Another fairly common problem occurs with system development projects. The client has an "itch" that needs scratching. Unfortunately, the client knows very little about the capability of software, and the programmers know very little about the user's operation. As they get into the project, they both begin learning, and the scope of the job begins to grow.

Planning was performed by a planning group. Although it is necessary in some environments, it can lead to disasters. Elsewhere in this handbook, I have told about the project manager who forgot the site preparation work on a construction job, which resulted in a $600,000 overspend on a job originally estimated at around $2 million. The cause—he planned it by himself, thus violating the rule that *the people who must do a job should participate in planning it.*

No one is in charge. This happens sometimes in organizations in which the project manager's role is not well defined or accepted by everyone in the organization. This can mean that no one person is really responsible for the project, so that things "fall through the cracks." When the project manager's role is weak, he or she may be given no approval authority over expenditures, thus resulting in a "blank check" for people in the organization. Managing a project by committee can also be a problem, especially if they have no skills in reaching consensus, etc.

Project estimates are best guesses, made without consulting historical data. It sometimes happens that there *is no historical data!* In many companies, good records of what actually happened in projects do not exist, so no one can refer to them for planning the next project. Or the reporting of labor hours was contaminated because salaried people do not report overtime (since they are not paid for it) and the current project was planned using the labor figures on the books.

Resource planning was inadequate. For example, no one bothered to check and see if a person with certain specialized skills would be available when needed in the project. Or a functional group is expected to do work for several projects, but no one noticed that the composite work load would require 300 percent more man-hours than were actually available in the department. Poor resource planning may very well be the most frequent cause of project failures.

People don't see themselves as working on one team. When work is divided up into different functional areas, individuals sometimes loose sight of the fact that the ultimate result requires the combining of all of the parts. They build walls around themselves, don't talk to or coordinate their efforts with members of other subgroups, and the result is chaos. A project manager has to work on team-building to avoid this.

People are constantly pulled off the project or reassigned with no regard for impact. This often happens because functional managers have no concern for projects. They are "rewarded" for making their functional departments run smoothly, not for achieving project objectives. Again, this is because people in organizations do not see the importance of project work or the project manager. Rewards must be consistent with what the organization wants to happen, since it is a psychological premise that *what is rewarded gets done.*

The project plan lacks detail. When a project is planned with too little detail, it is difficult to anticipate what kind of problems may develop. Further, it is hard to adequately manage resources, do proper estimates of time or costs, and develop workable schedules. Invariably this "broad-brush" approach to planning results in numerous conflicts and frequent changes, and creates interference with other projects being executed at the time. One caution, however: the opposite approach is not desirable either. A basic rule is that no project should be planned in more detail than can be managed. Clearly, a balance is required.

The project is not tracked against the plan. This seems inconceivable, but it happens. There are two general reasons. One is that the plan was a broad-brush plan, which had too little detail, so is not worth following, but then the team winds up having no control, since control can be exercised only by following a plan. The second problem is that a detailed plan is developed, but a problem develops during execution and people go into the panic mode and forget the plan. Again, they lose control. Planning should not be done just to satisfy some requirement that a plan exist. A plan that is not followed is useless.

Another thing that happens is that people take the attitude that the plan keeps changing, so they may as well abandon it. This is like saying that, since we have encountered several detours on our drive across the United States, we may as well throw out the map

- people don't work in teams
- the project is not tracked against the plan
- people lose sight of the original goal

and just "wing it." This is clearly nonsense, unless you don't care where you are at any given time.

People lose sight of the original goal. Engineers doing product development sometimes do this. They become so enamored of the technology that they forget they were supposed to be developing a product. Or they become perfectionists and waste time trying

241

to make it better than the specification calls for. Project managers must continuously monitor project activities and, if necessary, remind contributors of the purpose of their work.

Senior managers refuse to accept reality. Sometimes senior managers have an idea of what should be required to do a job based on knowledge of a previous job or some other factor. When a project manager turns in an estimate that is out of line with what the manager thinks it should be (with the estimate being higher, of course), the manager insists that it be reduced. If the project manager is coerced into agreeing to targets in this manner, the project will probably fail.

In line with this is the tendency to dictate performance, cost, schedule, and scope targets simultaneously, thus violating the rule that only three of the four can be pinned down—the fourth must be allowed to be what it is, as the four are inter-dependent.

Ballpark estimates become official targets. Sometimes a project manager is asked for a *ballpark* estimate, which is to be used for a "go/no-go" decision. Since people are only at the thinking stage, details are sketchy. The estimate is made based on that sketchy information. A decision is then made to do the job, but now more detail is available, and it turns out that the ballpark figure was way too low. At this point, however, the manager cannot go back to higher-ups and tell them that the job will cost more, since they were originally told a lower figure, and have come to think of that figure as *the* correct figure. So the ballpark becomes the target. To avoid this, the estimator needs to document all assumptions, make clear *in writing* that the estimate is a ballpark, with tolerances of x percent, so that it is clear from the beginning that the figures are subject to revision. (Of course, the ballpark may still become the target, but at least you have documented your original position.)

Section Six

Progress Payments

Chapter 18

Progress Payments and Earned-Value Analysis[*]

Some Background on Progress Payments

One of the major challenges facing all prime contractors is managing the risks of contract performance: the cost, schedule, and technical risks. With the progressively larger share of contract dollars now

---

[*] This chapter is from Chapter 1 and Chapter 5 of *Subcontract Project Management and Control: Progress Payments*, by Quentin W. Fleming and Quentin J. Fleming (Chicago: Probus, 1992). Reprinted by permission of the authors.

going outside of prime contractors down to their subcontracting base, many prime contractors are now looking outward at their suppliers to share some of the "glory" of the risks of contract performance.

There are a multitude of ways to minimize prime contractor performance risks; one of the more obvious is with the preparation of high-quality specifications for a program—system, performance, process, development, procurement—to mention just a few. Another key factor in risk management is the ability to define an airtight statement of work, both for internal budget performance and most particularly for subcontracted (external) supplier performance.

However, many times it is not possible to precisely define either a tight specification or even an adequate statement of work. In these cases the prime contractor may attempt to allot some of their own risks with subcontractors by selecting an appropriate contract type for the occasion—choosing from the two broad families of contract types, either a fixed-price or a cost-reimbursable contract.

The importance of selecting the appropriate type of contractual arrangement for a given subcontract is a critical one for all procurements. Selecting the type of contract should be decided by many factors, perhaps none more important to both parties than the amount of "cost risk" the buyer (the prime contractor) wishes to transfer to the seller (the subcontractor), and conversely, how much of the cost risk the seller is willing to assume. If the prime contractor is willing to retain the risks of cost growth, for whatever reason, then they will likely choose a cost-reimbursable type of contract.

If, however, it is the buyer's intent to transfer the maximum potential cost risk to the subcontractor, then some type of a fixed-price contract would be used, likely a Firm Fixed Price (FFP) subcontract.

Under a FFP-type subcontract, a supplier is obligated to assume the complete risk of any resulting cost growth and losses. This is the normal rule, but there is one very important condition, almost an exception, that should be understood. When the subcontract includes a "Progress Payment" clause in the arrangement, there is some likelihood that the cost risk factor *may* well have remained with the prime contractor and did not transfer to the supplier. And, although not a widely publicized fact, there is a history of—shall we say—"unfortunate cost experiences" (called losses) in the industry that have been the direct result of the poor management of progress payments. For example, one of the more obvious ways for a prime contractor to lose money is to make progress payments in advance of the supplier's physical performance, and then have the supplier close its doors.

Proper management of progress payments begins prior to making the subcontract award, and must continue throughout the life of the subcontract. A prime contractor that uses progress payments is particularly vulnerable if its buyers fail to understand the importance of doing the right things both prior to award and after subcontract award.

Therefore, with progress payments so very common in the government contracting business, and with prime contractors "encouraged" by the government to flow these financial arrangements downward to all their subcontracting team members, the subject and the associated risks of using progress payments must be clearly understood in order to best protect the interests of prime contractors and, of course, the United States Government.

Just What Are Progress Payments?

In their simplest form, progress payments may be viewed as a "loan," a temporary interest-free loan from a buyer (the prime contractor) to a seller (the subcontractor). They are based on costs incurred by the supplier in the performance of a specific order and are paid directly to the supplier as a stipulated and agreed-to percentage of the total costs incurred as defined in the Federal Acquisition Regulation (FAR). The supplier promises to "pay back" the temporary progress payment loan by (1) making contractual deliveries or by completing contractual line items, and (2) allocating some portion of the proceeds of the delivered unit price or completed line-item values to liquidate the loan, based on the established subcontract unit price of the articles or services delivered.

Note that there is an important distinction to be made between the "loan" value (the progress payment rate), and the repayment of the loan (the "liquidation" rate for the progress payments). Progress payments are paid as a percentage of *costs* incurred by a supplier—for example, 80 percent of the costs incurred. By contrast, the liquidation of the loan is based on a percentage of the unit *price* of article or service deliveries, which includes the supplier's fee. As such, the loan liquidations will include both the supplier's costs and the supplier's profit (i.e., the full subcontract unit values of the delivered articles).

Thus, the delivery of contract units and the completion of line items, once started, will result in a dramatic reduction of the outstanding progress payment loan. However, contract deliveries must first happen, and subcontractors must make physical progress in order to liquidate or repay the loan—this sometimes is the very heart of progress payment difficulties.

Progress payments are a form of contract financing to be used on *fixed-price* contracts. The payments cover the period that starts when a supplier begins to incur costs against an authorized order and continues to the time when the supplier is making unit deliveries or completing contractual tasks. This results in a supplier getting paid based on established unit or line-item task prices. The intent of progress payments is to prevent an "impairment" of the working capital of industrial suppliers doing business under United States Government contracts.

With this introduction in mind, we now turn to how the earned-value approach is used to control progress payments. For a complete treatment of progress payments, the interested reader is referred to the book by Fleming and Fleming, from which the material for this chapter was taken.

Progress Payments and the Earned-Value (C/SCSC) Concept

On January 7, 1991, Secretary of Defense Richard B. Cheney canceled the A-12 Avenger program. This has been reported to be the largest contract ever terminated by the DOD. Upwards of 9,000 employees immediately lost their jobs because of this single action.

Without debating the rightness or wrongness of the Cheney decision, to those of us who are interested in management control systems for government programs, particularly major programs, the A-12 Avenger cancellation provides a case study that will likely be discussed for years. To those of us who are specifically interested in the subset elements of "progress payments" and "earned-value" performance management, the A-12 incident provides a "lessons learned" opportunity of major importance.

The exact circumstances surrounding the A-12 will not be available to the general public for several years. It was what is called a secret SAR (special access required) program, which kept it out of the main monitoring processes of the DOD and certainly from the general public. Nevertheless, enough public information has surfaced for us to draw certain conclusions.

The prime contracts were awarded on January 13, 1988, under a fixed-price incentive contractual arrangement, which contained a target price of $4.379 billion, a target cost of $3.981 billion, and a ceiling price of $4.777 billion.[1] Full compliance with the Cost/Schedule Control Systems Criteria (C/SCSC) and periodic Cost Performance Reports (CPR) were required from the two prime contractors, McDonnell Douglas of St. Louis, and General Dynamics of Fort Worth.[2] Progress payments were included in the fixed price contractual arrangement.

It has been acknowledged by reliable DOD sources that the C/SCSC management control systems were implemented properly, and were functioning well at both the principal contractors.[3] But as early as April 10, 1991 (some 90 days after cancellation), it was reported that the government was demanding a return of $1.35 billion in "overpayments" made to the two principal contractors.[4] And by June 8, 1991 (five months after cancellation), the two prime contractors had filed a 78-page lawsuit against the government, arguing they are entitled to keep the questioned overpayment of funds.[5] Stay tuned—this saga will be continued.

Final settlement of this "major difference of opinion" between the United States government and two of its largest contractors will likely take years to settle in the courts. However, if it is generally acknowledged that the C/SCSC management control systems were working well with both of the prime contractors, and there was an overpayment of one-third of the total program's target costs only partway through the contractual period, then one can only conclude that the C/SCSC administrators appear not to have been communicating well with the progress payment administrators! Thus, contractor progress payments would appear not to have been linked with earned-value (C/SCSC) performance measurement. Time will tell.

This chapter will not focus on the A-12 program cancellation. The A-12 merely provides us with a case study, a role model, of what can go wrong, and perhaps some examples of future practices we will want to do differently ourselves, if at all possible.

Rather, in this chapter we will want to address four basic subjects in a generic sense, attempting to "link" the activities of progress payments with the earned-value performance measurement concept.

1. A brief overview of the "earned-value" (C/SCSC) performance measurement concept;

2. Relating progress payment data to earned-value performance measurement data when full C/SCSC is formally imposed on the subcontractor;

3. Relating progress payment data to performance on firm-fixed-price (FFP) type subcontracts when there is no formal C/SCSC imposed on the supplier;

4. Methods (the formula) used by earned-value (C/SCSC) practitioners to forecast an independent estimate of costs at completion (EAC), based on the actual cost and schedule performance of the subcontractor.

The Earned-Value (C/SCSC) Concept in a Nutshell

In spite of the title, the earned-value performance measurement concept is a complex subject that is difficult to present in a "nutshell." One of the best introductions to the theory of the earned-value concept comes from one of its founders, in a recent article that he wrote after retirement from the government:

Since 1967, the DOD has employed the Cost/Schedule Control Systems Criteria (C/SCSC) as a means to ensure that major contractor's internal management systems are sound and can provide government program managers with reliable, objective cost performance information for use in management decision making. The "criteria approach" allows contractors to adopt the systems and controls of their own choosing, provided those systems can satisfy the criteria. Compliance is determined by government teams, which review the systems in operation after contract award.

The C/SCSC require that a contractor establish an integrated cost and schedule baseline plan against which actual performance on the contract can be compared. Performance must be measured as objectively as possible, based on positive indicators of physical accomplishment rather than on subjective estimates or amounts of money spent. Budget values are assigned to scheduled increments of work to form the performance measurement baseline (PMB).

In order to measure contract performance, budgets for all work on the contract must sum to the contract target cost (CTC) so that each increment of work is assigned a value (budget) that is relational to the contract value. When an increment of work is done, its value is earned; hence the term *earned value*. By maintaining the budgetary relationship to contract target cost, variances from the budget baseline reflect ongoing contract cost performance.[6]

To properly put the earned-value performance measurement concept into historical perspective, we must go back in time some three decades and trace the evolution of Cost/Schedule Control Systems Criteria (C/SCSC) from its two ancestors: PERT/Time and PERT/Cost.

The Program Evaluation and Review Technique (PERT) was introduced by the United States Navy in 1957 to support the development of its Polaris missile program. PERT was a technique that attempted to simulate the necessary work to develop the Polaris missile by creating a logic network of dependent sequential events. Its purpose was threefold: to plan the required effort, to schedule the work, and to predict the likelihood of accomplishing the objectives of the program within a given time frame. The initial focus of PERT was on the management of time, and predicting the probability of program success.

There was great excitement surrounding the new PERT program management concept. Unfortunately, the technique's successes fell far short of the proponents' expectations. Part (perhaps most) of the difficulty with PERT was not with the concept itself, but rather with the computers of the time. Both computer hardware and software were not up to the required challenges in 1957. Computers were scarce, and PERT network processing had to compete with the processing of the company's payroll, and somehow the company payroll always won. But also the software programs, evolving initially out of simple linear network concepts, just could not provide the needed flexibility to support the program management requirements at the time.

While the PERT planning, scheduling, and probability forecasting concepts have survived to this day, the technique for use as a program management tool initially "suffocated" a few short years after its introduction. The technique was too rigid for practical applications with the computer hardware and software available at the time. And there was also the problem of the overzealous government mandate to use PERT. Industry management rightfully resented being told what tools they must use in the management of their contracts.

Then, before PERT was accepted by program management in industry, the United States Air Force came up with an extension of PERT by adding resource estimates to the logic networks. "PERT/Cost" was thus born in 1962, and just plain "PERT" was thereafter known as "PERT/Time." Needless to say, if PERT/Time as a management technique was too rigid for practical applications at the time, PERT/Cost with the added dimension of resources only exacerbated the problem. PERT/Cost as a management control tool had a lifetime of perhaps two years.

What was significant about PERT/Cost, however, was not the technique itself, but rather what evolved from it. The "earned-value" measurement concept was first introduced to industry when the government issued their *Supplement No. 1 to DOD and NASA Guide, PERT/Cost Output Reports*, in March 1963, in which they provided industry with a simple definition of the earned-value concept:

> VALUE (Work Performed to Date): The total planned cost for work completed within the summary item.[7]

Thus, instead of relating cost plans to cost actuals, which historically had been the custom, PERT/Cost related the "VALUE" of work performed against the cost actuals, to determine the utility/benefits from the funds spent. What was *physically accomplished* for what was *actually spent* was a simple but fundamentally important new concept in program management. Hence, the earned-value concept was first introduced in 1963, but had to wait until the issuance of the formal C/SCS Criteria to have its full and lasting impact on American industry.

For various reasons the United States Air Force gave up on the PERT/Cost technique in the mid-1960s, but correctly held on to the "earned-value" concept. When the Department of Defense formally issued their Cost/Schedule Control Systems Criteria (C/SCSC) in 1967, the earned-value concept was solidly contained therein. With the subsequent adoption of these same identical criteria by the Department of Energy in 1975, and the re-affirmation of the criteria by the Department of Defense in their major 1991 defense acquisition policy statement, the earned-value concept of cost and schedule management is firmly established in the United States government acquisition process.[8]

A detailed discussion of the 35 specific criteria contained in the C/SCSC is beyond the limited scope of this book. There are full textbooks, and week-long seminars, and practitioners/consultants available to cover these matters should one have the requirement or even the inclination. Rather, we will merely attempt to summarize some of the more significant features of the concept so as to be able to relate the earned-value concept to our primary subject: progress payments to fixed price subcontractors.[9]

The C/SCSC are divided into five logical groupings, which contain the 35 criteria:

1. **Organization** *(five criteria)*: To define the required contractual effort with use of a work breakdown structure (WBS), to assign the responsibilities for performance of the work to specific organizational components (i.e., the organizational breakdown structure (OBS)), and to manage the work with use of a single "integrated" contractor management control system.

2. **Planning and Budgeting** *(11 criteria)*: To establish and maintain a performance measurement baseline (PMB) for the planning and control of the authorized contractual work.

3. **Accounting** *(seven criteria)*: To accumulate the actual costs of work performed (ACWP) and materials consumed in a manner that allows for its comparison with the actual performance measurement (BCWP).

4. **Analysis** *(six criteria)*: To determine the earned value, to analyze both cost variances (CV) and schedule variances (SV), and to develop reliable estimates of the total costs at completion (EAC).

5. **Revisions and Access to Data** *(six criteria)*: To incorporate changes to the controlled performance measurement baseline (PMB) as required, and to allow appropriate government representatives to have access to contract data for determining C/SCS Criteria compliance.

We will briefly discuss some of the critical elements of these five criteria groupings to provide a quick overview of the earned-value concept. Each of the acronyms used above will be defined in the discussion that follows.

Criteria Group 1

The five criteria required by the first group covering organization can best be illustrated by a review of the diagram in Figure 18.1. Criterion one (#1a) requires the use of a work breakdown structure (WBS) to define the required effort, whether it be a contract, sub-

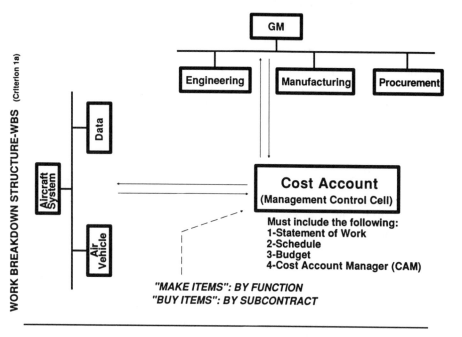

Figure 18.1. Work Breakdown and Organizational Breakdown Structures.

ORGANIZATIONAL BREAKDOWN STRUCTURE-OBS (Criterion 1b)

contract, a company-funded internal project, etc. The WBS approach allows program management to comprehensively define and then to perform a given contract within the maze of a company's functional organization. The use of a WBS to define the program is illustrated in Figure 18.1, at the extreme left side.

The second criterion in group 1 (#1b) requires assignment of the defined WBS work tasks to the organizational breakdown structure (OBS) for performance. This concept is illustrated in the upper portion of Figure 18.1. Internal functional organizations (OBS) will perform the contract tasks as defined by the WBS.

The third criterion (#1c) requires the integration of the contractor's management control functions with each other, and with the defined WBS and OBS elements. This requirement is achieved by the creation of "management control cells," which are referred to in

C/SCSC as Cost Accounts, and are displayed in Figure 18.1. Tasks
that are "make items" (work to be performed internally by the contractor) must be identifiable to both the WBS and OBS. Likewise,
"subcontracts" must also be identifiable to both the WBS and OBS.
Thus all of the hundreds (or more) of these self-contained management control cells (Cost Accounts) in C/SCSC must be relatable to
either the WBS to comply with the precise language of the prime
contract's statement of work, or by the OBS to satisfy the requirements of internal functional management within a given company.

Each management control cell (cost account) must have the following four elements to maintain the integrity of the management
control unit, and for the performance measurement of data contained therein: (1) a statement of work for the cell; (2) a time frame
or schedule for the cell; (3) a budget of financial resources; (4) a
responsible manager, typically referred to as the Cost Account Manager (CAM). The cost account concept or management control cell is
fundamental to C/SCSC and is displayed in Figure 18.1, in the
lower right corner.

Criteria Group 2

The eleven criteria contained in the second group covering *planning*
and *budgeting* require the formation of a measurement baseline
against which the supplier's performance may be measured. This
requirement can be illustrated by our reviewing Figure 18.2, the
Performance Measurement Baseline (PMB). There are 12 specific
components to what is called the C/SCSC Performance Measurement Baseline, and to follow the discussion, one must have some
understanding of what is meant by each of the elements contained
in the baseline. Therefore, these twelve PMB elements must be defined for us, relatable by number to the elements displayed in Figure 18.2.

1. **Contract (or Subcontract) Target Price (CTP):** The negotiated
   estimated cost plus profit or fee for the contract or subcontract.

2. **Fee/Margin/Profit:** The excess in the amount realized from the
   sale of goods, minus the cost of goods.

3. **Contract (or Subcontract) Budget Base (CBB):** The negotiated contract cost plus the contractor's (or subcontractor's) estimated cost of authorized but unpriced work.

4. **Contract (or Subcontract) Target Cost (CTC):** The negotiated cost for the original definitized contract and all contractual

Figure 18.2. The Performance Measurement Baseline (PMB).

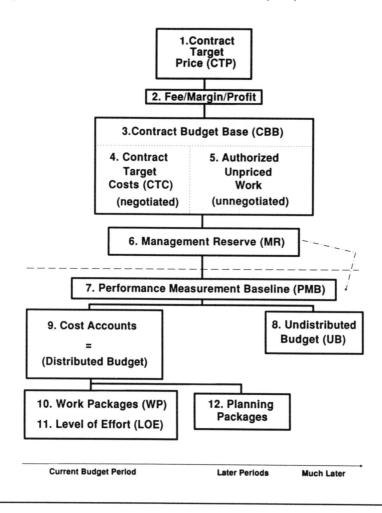

changes that have been definitized, but excluding the estimated cost of any authorized, unpriced changes.

5. **Authorized Unpriced Work**: The effort for which definitized contract costs have not been agreed to, but for which written authorization has been received by the contractor/subcontractor.

6. **Management Reserve (MR)**: A portion of the Contract Budget Base that is held for management control purposes by a contractor to cover the expense of "unanticipated" program requirements. MR is not initially a part of the Performance Measurement Baseline (PMB), but is expected to be consumed as PMB prior to completing a contract. Any MR not consumed at program completion becomes pure profit, #2 above, or profit and some portion returned to the buying customer under an incentive-type arrangement.

7. **Performance Measurement Baseline (PMB)**: The time-phased budget plan against which project performance is measured. It is formed by the summation of budgets assigned to scheduled cost accounts and their applicable indirect budgets. For future effort that is not currently planned to the cost account level, the Performance Measurement Baseline also includes those budgets assigned to higher-level WBS elements. The PMB normally equals the contract budget base, less management reserve.

8. **Undistributed Budget (UB):** Budget applicable to contract effort that has not yet been identified to WBS elements at or below the lowest level of reporting to the government or prime contractor.

9. **Cost Account (CA):** A natural intersection point between the work breakdown structure (WBS) and the organizational breakdown structure (OBS), at which functional management responsibility for the work is assigned, and where actual direct labor, material, and other direct costs are compared with earned value for management control purposes. Cost accounts are the focal point of cost/schedule control.

10. **Work Packages (WP):** Detailed short-span jobs or material items, identified by the contractor for accomplishing work re-

quired to complete a contract. Work packages are discrete tasks that have specific end products or end results.

11. **Level of Effort (LOE):** Work that does not result in a final product (e.g., liaison, coordination, follow-up, or other support activities) and which cannot be effectively associated with a definable end-product process result. It is measured only in terms of resources actually consumed within a given time period.

12. **Planning Package:** A logical aggregation of far-term work within a cost account that can be identified and budgeted but not yet defined into work packages. Planning packages are identified during the initial baseline planning to establish the time phasing of the major activities within a cost account and the quantity of the resources required for their performance. Planning packages are put into work packages consistent with the "rolling wave" scheduling concept prior to the performance of the work.

It sometimes comes as a surprise to some that the C/SCSC Performance Measurement Baseline (PMB) represents a value less than the total contract or subcontract amount. However, this fact is true only for the initial PMB. Profit or fee is not intended to be used in the performance of the contract, or else the result will be zero profit to the contractor or subcontractor. By contrast, management reserve (MR) is expected to be consumed during contractual performance, and when it is needed, MR is shifted into the PMB. Any management reserve remaining at the end of the contract is used to offset unfavorable variances, or may represent contract underrun, and/or profit.

Criteria Group 3

The seven criteria in the *accounting* group require that both cost actuals and schedule performance be relatable in the same time period with the "earned-value" achievement, or what PERT/Cost called simply "value." Important point: by definition in C/SCSC, a cost variance (CV) is the difference between the earned value achieved and the cost actuals for the same period. A schedule variance (SV) is the difference between the value of the scheduled work for the period and the earned value achieved in the same period. Let us discuss this simple but fundamental concept with a review of the data displayed in Figures 18.3 and 18.4.

Figure 18.3 presents an imaginary four-year, $100-million contract, using the "conventional" cost control method. The plan calls for the expenditure of exactly $25 million each year for the four years. At the end of exactly two years of performance we find we have spent $40 million, compared with our plan, which called for the expenditure of $50 million. How are we doing? Truthful answer: we really do not know, using the "conventional" planned costs versus actual cost versus actual costs method.

Figure 18.3. Conventional Cost Control.

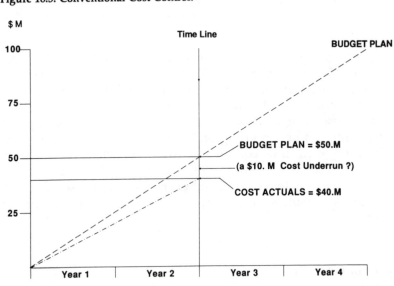

An optimist might look at the data in Figure 18.3 and say we have accomplished $50 million of work, and we have spent only

$40 million in actual costs, so therefore we have underrun our costs to date by some $10 million. And remember, most program managers and senior executives and CEOs are optimists by their very nature!

Or, a pessimist might look at this same chart and conclude that we have completed some $40 million of our planned work and have spent exactly $40 million in actual costs, so therefore we are the equivalent of $10 million of work behind our schedule. But on the other hand, our cost performance is just fine!

In reality, we cannot tell how well or poorly we are doing by simply using the "conventional" cost control method of comparing planned cost expenditures with actual cost expenditures. Our conclusions can be, and will likely be, most deceiving. We cannot tell if we have overrun or underrun our costs or are ahead or behind in our schedule using the traditional methods of cost management. And our being an optimist or a pessimist or a realist will not improve the process. We need to know what physical work we have accomplished against our physical work plan, for the actual dollars we have spent in a given time frame. We need "earned-value" performance measurement to be able to make an objective assessment of our program accomplishments.

Going on to Figure 18.4, we have added the critical third dimension of "earned-value" performance measurement, and are shocked at the results. We have only accomplished $25 million in physical work. Therefore, since we have spent $40 million, we are in fact $15 million overrun in costs. And to add to our distress, of the $50 million in work we had planned to accomplish in the initial two years, only $25 million has been accomplished. We are thus $25 million in equivalent work behind schedule, or stated another way, we are one year behind schedule!

By equating our cost dollars spent to our earned value, we know (sometimes painfully) exactly how well or poorly we are doing in our cost performance. By equating the planned schedule with earned value we know (sometimes painfully) how much of the authorized contract work we have accomplished against our own schedule. Earned-value performance measurement is "objective" measurement. It takes the guesswork out of the management of cost and schedule management of contracts or subcontracts. And per-

Figure 18.4. "Earned-Value" Performance Measurement.

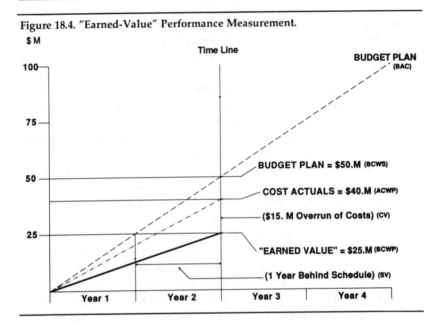

haps of greatest importance, the supplier's actual cost and schedule performance can be used to intelligently forecast the final estimate of costs to complete the effort, and the necessary time to complete the effort.

In order to follow the discussion that will be covered in the final section in this chapter on using C/SCSC performance data to predict the final outcome, we will need to master some of the C/SCSC jargon. However, if we cover these very specific terms and immediately relate them to the data contained in Figure 18.4, perhaps we can minimize the pain. Nine definitions of specific C/SCSC terms are needed:

Budgeted Cost for Work Scheduled (BCWS): The sum of the budgets for all work packages scheduled to be accomplished (including in-process work packages) plus the amount of level of effort scheduled to be accomplished within a given time period.

(The BCWS is nothing more than the "plan" against which contractor performance will be measured. In Figure 18.4 the plan or BCWS through year 2 was $50 million.)

Budgeted Cost for Work Performed (BCWP): The sum of the budgets for completed work packages and completed portions of open work packages, plus the appropriate portion of the budgets for level of effort. Also known as "Earned Value."

(The BCWP is the "earned value," the physical value of the work done at a given point in time. In Figure 18.4 the BCWP, or earned value, at year 2 was $25 million.)

Actual Costs of Work Performed (ACWP): The costs actually incurred and recorded in accomplishing the work performed within a given time period.

(The ACWP is the "actual costs" for a given period. The ACWP through year 2 was $40 million.)

Cost Variance (CV): The numerical difference between earned value (BCWP) and actual costs (ACWP).

(The CV is the difference between the earned value (BCWP), less the actual costs (ACWP). In Figure 18.4 the earned value (BCWP) is $25 million, less actual costs (ACWP) of $40 million, for a CV of $15 million.)

Note the important difference in C/SCSC performance measurement: There is no comparison of the plan (BCWS) with the actual costs (ACWP), as with the "conventional cost" method.

Schedule Variance (SV): The numerical difference between earned value (BCWP) and the budget plan (BCWS).

(The SV is the difference between what was scheduled to be done, BCWS, and what was accomplished, or the earned value (BCWP). In Figure 18.4 the BCWP is $25 million, less the BCWS of $50 million, for an SV of $25 million).

Budget at Completion (BAC): The sum of all budgets (BCWS) allocated to the contract. It is synonymous with the term Performance Measurement Baseline (PMB).

(The BAC is important as a comparison with the estimate at completion, which will take place during the period of performance. In Figure 18.4 the BAC is $100 million at the end of four years, synonymous with the BCWS in this case.)

Estimate at Completion (EAC): A value periodically developed to represent a realistic appraisal of the final cost to complete an effort. It is the sum of direct and indirect costs to date, plus the estimate of costs for all authorized work remaining. EAC = ACWP + the Estimate to Complete.

(Thus, whenever we apply sound business practices to ourselves, or with progress payment administration, or with C/SCSC performance measurement, periodically we will want to make an estimate of what it will take to complete a given job. In Figure 18.4 there is no EAC forecasted by the supplier. However, with a CV of $15 million and an SV of $25 million only halfway through the contract period, a realistic EAC is definitely in order.)

Cost Performance Index (CPI): The cost efficiency factor achieved by relating earned-value (BCWP) performance to the actual dollars spent (ACWP).

(The CPI is the critical indicator of program performance using the earned-value technique. The CPI is derived by dividing the earned-value BCWP ($25M) performance by the dollars actually spent (ACWP) ($40M), which provides the cost efficiency factor for work accomplished after two years. The result of $25M divided by $40M provides an efficiency factor of 62.5 percent. Stated another way, for every dollar spent to date, this program achieved a benefit of only 62.5 cents!)

Schedule Performance Index (SPI): The schedule efficiency factor achieved by relating earned value (BCWP) against the scheduled work (BCWS).

(The SPI is a critical corollary index to the CPI, and is often used in conjunction with the CPI to forecast the final outcome. The SPI is derived by dividing the earned-value (BCWP) performance of $25M by the scheduled work BCWS ($50M), which provides the schedule efficiency factor for work after two years. The result of $25M divided by $50M equates to a factor of only 50 percent. Stated another way, for every dollar of equivalent work planned, this contract only accomplished one-half of it.)

(Thus, one must conclude that the work initially scheduled to be done in years 1 and 2, will now be performed in an extended contract period into year 5 or even 6. And everyone knows the simple truth that work done in later periods likely will cost more to accomplish because of inflation.)

With these nine definitions, related back to the data contained in Figure 18.4, we will be in a position to better understand the forecasting techniques in the final EAC section below.

Criteria Group 4

The *analysis* section contains six criteria that require the contractor or subcontractor to make an assessment of what has occurred, with its cost and schedule performance to date. Most important, however, this criteria group requires the supplier to analyze the cost and schedule performance to date, and then to estimate the cost and schedule requirements necessary to complete the effort, to forecast the EAC. Estimates at Completion will be covered thoroughly later in this chapter.

Criteria Group 5

The *revisions and access to data* section contains six criteria, which require that the supplier maintain the performance measurement baseline (PMB) throughout the life of the contract by incorporating all new work into the PMB in a timely manner. Obviously, the maintenance of the baseline is vital to the integrity of any performance measurement system. Also, this group requires the contractor to provide access of performance data to the customer's representatives in order to verify strict compliance with the criteria.

Since the issuance of the C/SCS Criteria by the DOD in 1967, the application of the concept has been intentionally limited to only those contracts in which the customer (the buyer) has retained the risks of cost growth (i.e., on cost or incentive-type contracts/subcontracts). The dollar thresholds for formal C/SCSC implementation will vary from period to period and will be set by the buying customer, but are currently generally imposed at $50 million on prime contracts ($25 million for subcontracts) for developmental-type work, and $160 million ($60 million for subcontracts) for production-type efforts. Any full application of C/SCSC will require a periodic report (typically monthly) called the Cost Performance Report (CPR).

The lesser Cost/Schedule Status Reports (C/SSR) or the Cost Performance Reports/No Criteria (CPR/NC) are now generally set on smaller contracts at $5 million in contract value and a minimum of 12 months in program duration. However, program and subcontract management should weigh the risk factors involved in a given effort and decide the earned-value applications on a case-by-case basis.

This overview was out of necessity a very limited discussion of a very large subject. One last question needs to be addressed: Does the earned-value (C/SCSC) performance measurement concept really work, or is it just another government requirement? To best answer this question we should review the results of an impressive Department of Defense study.

Covering the period from 1977 to 1990, over 400 DOD contracts in which formal C/SCSC was implemented were studied. The results of the analysis are most impressive, and without exception their findings were consistent for all of the 400-plus contracts monitored.

Once the C/SCSC performance measurement baseline (PMB) is in place and at least 15 percent of the planned work has been performed, the following conclusions can be made on the future performance of a given program:

☞ The overrun at completion will not be less than the overrun to date.

☞ The percent overrun at completion will be greater than the percent overrun to date.

☞ The conclusion: You can't recover.

☞ Who says? More than 400 major DOD contracts since 1977.

☞ Why? If you underestimated in the near term, there is no hope that you will do better on the far-term planning.[10]

In closing on this section covering the basics of the earned-value concept, we should cover the objectives sought to be gained through the use of the performance measurement concept to monitor and manage our contractors/subcontractors. Once again, going to one of the originators of the concept who did so much to implement the technique during his tenure with the DOD and later at the DOE, he summarizes the four objectives we can obtain from employing the earned-value (C/SCSC) performance measurement concept on our programs:

☞ Sound Contractor Systems

☞ Reliable, Auditable Data

☞ Objective Performance Measurement

☞ No Surprises[11]

Four simple programmatic objectives, not always easy to obtain.

Comparing Progress Payments Data with C/SCSC Cost Performance Report (CPR) Data

When comparing the cost data contained on progress payment invoices with the cost data contained in the formal C/SCSC Cost Performance Report (CPR), or the lesser C/SSR or CPR/NC, there are two issues that require a reconciliation when there are differences reported from a contractor or subcontractor, and the reporting of differences will likely be the norm, not the exception:

1. The comparison of the *cost actuals* (ACWP) reported to date—between those contained in the progress payment invoice, versus the cost actuals reflected on the CPR;

2. The comparison of the *estimates to complete* (EAC) the effort—between those forecasted on the progress payment invoice, versus those forecasted on the CPR.

For purposes of this discussion we will consider the three distinct C/SCSC cost and schedule performance reports (CPR, C/SSR, CPR/NC) as being identical for the purpose of reviewing the data contained therein. Any generic differences in these three reports deal with other matters, not the actual costs (ACWP) or the estimate at completion (EAC) contained in these reports.

One would expect that there is, or that there should be, some direct relationship between what a contractor/subcontractor reports as their cost actuals position when they submit progress payment invoices, versus what they may reflect on other cost reports (e.g., a CPR). After all, the cost data does come from the same supplier, reporting their cost status from a single accounting system.

However, in practice, it is not unusual to receive multiple cost reports from a supplier reflecting different financial actuals for the same reporting period. Anytime this happens, it is incumbent on the buyer to request a reconciliation from the subcontractor, requiring an explanation of any differences in the cost reports.

The culprit in such discrepancies can be any of several factors that can make the data contained in any of these cost reports unique. These factors are:

1. Different cutoff dates for the reports, or the data contained or reported therein. In some cases the reports are to reflect a specific accounting closure date, but at other times the date will reflect the date of report submittal. Not infrequently, progress payment closure dates will have a different cutoff date from the general ledger closure date.

2. Cost data only (which excludes fee or profit), versus "price" data, which will include some estimate of earned fee or profit on lower-tier subcontracts. Often there are distinct professional differences of opinion between a buyer and seller as to how much profit or fee a given supplier will likely have earned on the effort at a given point in time.

3. Progress payments to lower-tier suppliers are included or excluded in the cost actuals being reported.

4. Negotiated statement of work versus unnegotiated statement of work, that is, changes. Unnegotiated work will often be placed

in several categories: (1) authorized, priced, and proposed; (2) authorized but unpriced, and unproposed effort; (3) unauthorized and still under discussion, etc. Not infrequently, there are legitimate differences of opinion between buyer and seller as to the correct value of the "yet to be negotiated" work.

5. The projected estimated supplier costs at completion, in absolute overrun or underrun terms, with the cost sharing impact it may have on a supplier's earned profits under an incentive-type contractual arrangement.

6. Termination liability projections at any given point in time, which will include either a supplier's open commitments or their expenditures only. Remember, small businesses may include accounting accruals as cost actuals for purposes of requesting progress payments. Large businesses must actually pay the bills in order for them to quality as actual expenditures.

7. Materials purchased (e.g., raw stock, nuts, bolts, chemicals, etc.), received, and placed directly into inventory, but not yet charged to work in process, the costs of which may or may not be incorporated into the cost actuals reported.

There are doubtless additional factors, all legitimate reasons, that may cause differences in the reported data between the progress payment requests (SF1443) and the C/SCSC cost performance reports (CPR). These seven items are not intended to be all-inclusive.

What all this means to the buyer and the seller is that they should insist that those who prepare such cost reports invest in a few choice narrative words in order to better describe the cost reports they send out, or in other words to set out what assumptions they may have made when the data was submitted. This is particularly true when there are similar financial terms being used in multiple cost reports, which can have different meanings to the practitioners. However, in every case when there are different values reported, the buyer has a programmatic responsibility to understand the reasons causing these discrepancies, prior to authorizing funds for the progress payment invoice.

Comparing Progress Payment Data Without the Benefit of Formal C/SCSC on Firm-Fixed-Price (FFP) Subcontracts

How do you get earned-value performance measurement on sub-contracts that do not have formal C/SCSC requirements imposed? One approach would be to universally impose full C/SCSC requirements on all contracts and subcontracts that are funded by the U.S. government. One individual from the government has suggested exactly that in his timely article on the subject of managing contractor progress payments. Note, in this quote he uses the term "flexible" to mean cost or incentive type contracts, and "inflexible" to refer to firm-fixed-price (FFP) type contracts.

> The DOD would be well advised to insist validated cost/schedule procedures be implemented on all large dollar contracts—flexible and inflexible, prime and sub-contracts—that require payment reviews by the government. Validated cost/schedule reporting will ensure proper program controls and provide the government with a more effective and efficient method of conducting government payment reviews.[12]

This gentleman makes the point that since there are hundreds of contractors that have fully validated C/SCSC systems, and since it is important to connect the approval of progress payments with the physical performance of a contractor, why not extend C/SCSC to all programs that are funded by the government. These are certainly valid points, but they are not recommended for a number of reasons.

In the first place, although there are currently over 200 actively validated C/SCSC management control systems in the United States, that number represents only a small fraction of the total contracts and subcontracts that are covered by government progress payments. Most of the firm-fixed-price FFP type contracts and subcontracts would not be impacted by such an edict, because most suppliers do *not* possess a validated C/SCSC management control system. Two hundred approved management systems out of several thousand suppliers is but a small percentage of the total.

But of greater significance, extending full C/SCSC to all programs that have progress payments would simply increase the costs of the government's procurement of major systems. The full application of formal C/SCSC with their 35 specific criteria have too much "nonvalue" added requirements to be universally and indiscriminately applied to all programs that have progress payments. Full C/SCSC applications should be limited to cost or incentive type contracts, with their inherent cost risks, which can benefit from having an early-warning monitoring system.

In 1967, when C/SCSC was first introduced, there was some confusion as to whether or not the criteria should be imposed on other than cost or incentive type contracts/subcontracts. It was decided at that time to limit the formal application of C/SCSC to those efforts where the risk of cost growth was on the buyer (i.e., on cost or incentive type contracts). That principle is still valid today. There are better, less costly ways to achieve the same goal, that of linking progress payment approvals to the physical performance of the supplier requesting payment.

When a supplier requests progress payments and the buyer is prudent enough to require the creation and monthly submittal of a Gantt chart, the prime contractor (buyer) has all that is needed to employ at least a modified version of the earned-value concept. Even a modified earned-value approach can be significant when monitoring fixed-price suppliers, who traditionally have refused to allow any performance monitoring by prime contractors.

It is important to require a Gantt chart for a project, which should be prepared by the subcontractor, who is required to list all of the planned tasks necessary to perform the order. Each of the listed tasks must receive a weighted value, the sum of which must add up to 100 percent of the purchase order price. What the Gantt chart with weighted values provides is in effect a simple form of an "earned-value" plan, which in the C/SCSC vernacular is referred to as the Budgeted Costs for Work Scheduled (BCWS). With the supplier's own plan, we can quantify each task with its value into a time frame to form a cumulative percentage curve. In Figure 18.5 we have illustrated the approach with data that would be supplied

Figure 18.5. Establishing the BCWS.

Hypothetical Engines, Inc.

Item#	Task	%	J	F	M	A	M	J	J	A	S	O	N	D	J	F	M	A	M	J	J	A	S	O	N	D	J	F	M	A	M	J
1	Des.Mod.	5	1	2	2																											
3	Qual.Test	5				5																										
5	Pur.Mat.	20				4	4	4	4	4																						
7	Fab.Parts	10								2	2	2	2	2																		
9	Comm.Assy	12										1	2	1	2	1	2	1	2													
11	#1	4																		2	2											
12	#2	4																			2	2										
13	#3	4																				2	2									
14	#4	4																					2	2								
15	#5	4																						2	2							
16	#6	4																							2	2						
17	#7	4																								2	2					
18	#8	4																									2	2				
19	#9	4																										2	2			
20	#10	4																											2	2		
21	#11	4																												2	2	
22	#12	4																													2	2
BCWS	Month%	100	1	2	2	9	4	4	4	6	2	3	4	3	2	1	2	1	2	2	4	4	4	4	4	4	4	4	4	4	4	2
BCWS	Cum.%		1	3	5	14	18	22	26	32	34	37	41	44	46	47	49	50	52	54	58	62	66	70	74	78	82	86	90	94	98	100

by a subcontractor to quantify its own performance plan, its own BCWS, monthly and cumulative.

Each month, as the supplier reports its actual performance against its own Gantt schedule, it must report a percentage completion against the plan. Suppose that in previous reports the supplier was reporting 28 percent complete as of October 1991, and later, 34 percent as of January 1992. This compares unfavorably with our assessment of its own data in Figure 18-5 To best illustrate what this supplier is reporting to us in its schedule performance, we should lay out the data as follows:

	Oct. 1991	Jan. 1992
BCWS Plan (from Figure 18.5)	37%	46%
BCWP Performance (from previous reports)	28%	34%
Schedule Variance Position	–9%	–12%
SPI (BCWP divided by BCWS)	76%	74%

In Figure 18-6 is shown what the subcontractor originally planned to do. One can immediately see that with the passage of time, this supplier is getting progressively further behind in accomplishing the work it set out to do in its own plan. And by measuring its schedule position with earned-value performance indices, we can quantify precisely how well or poorly it is doing. Their schedule performance index (SPI) went down from 76 percent of accomplishing planned work in October 1991, to 74 percent 90 days later.

What this percent complete estimate provides is a sort of modified earned value, or BCWP (Budgeted Costs for Work Performed) in the C/SCSC terminology. The difference between their planned BCWS versus what they accomplished in their BCWP, provides the schedule performance (SV) position for a subcontractor. It tells the prime contractor whether the supplier is accomplishing the work they had set out to do, and in a timely fashion. The performance of Hypothetical Engines is not going well, 12 months into a 30-month effort.

Figure 18.6. A Gantt Chart to Support Progress Payment Evaluations.

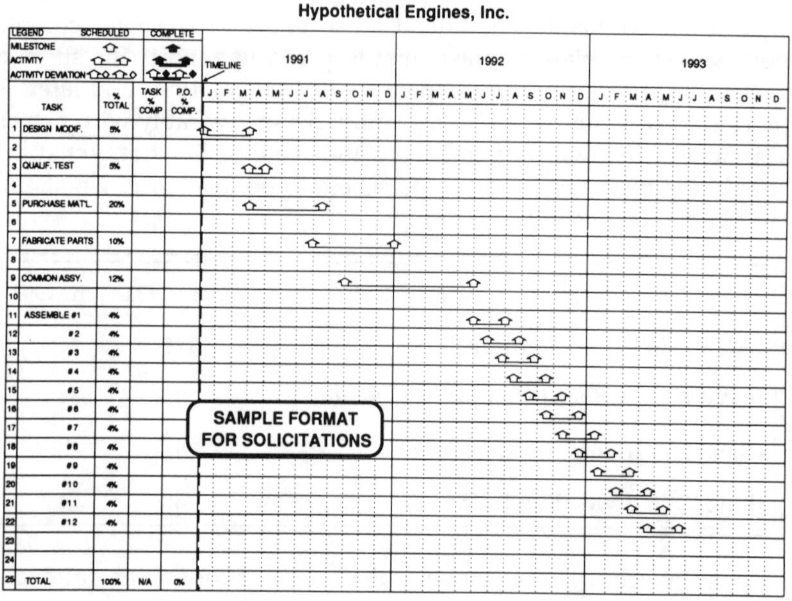

Hypothetical Engines, Inc.

Now to relate the earned value (percent complete) to the costs this supplier is experiencing. Each month as the supplier submits their request for progress payments, they must complete a SF1443 invoice form. Line 12a of the SF1443 contains their total actual costs incurred, cumulative to date. The amount listed on line 12a is equivalent to the actual cost values in C/SCSC, or what is called the ACWP, actual costs for work performed. When you relate the ACWP to the earned value (BCWP), you have the cost variance (CV) for performance by a supplier. With this information one can deduce the cost performance efficiency factor for the supplier (BCWP divided by the ACWP) to determine how much a supplier has earned for every dollar it has spent. If the supplier spends $1.00, but only accomplishes $0.85 in earned value, one should watch the supplier closely. It could be heading for a loss which will require the application of the loss ratio to all progress payments made once the total "projected" costs penetrates the subcontract price value.

Thus, by requiring that a subcontractor put in place a few elementary cost and schedule plans prior to a subcontract award, a buyer can employ a simple but effective earned-value performance measurement concept. With this, the performance of even firm-fixed-price suppliers may be monitored during the life of the subcontracts, and provide a linkage between progress payment approvals and earned-value measurement.

Using C/SCSC Performance Indices to Predict the EAC

The best Estimate at Completion (EAC) forecast for a given program is typically referred to as a "bottoms-up" or "grass-roots" EAC. Here, each of the remaining tasks to be worked is examined by the very persons who will perform the tasks, and a detailed estimate to complete all of the work is prepared. However, to do a legitimate bottoms-up EAC takes a lot of program resources to accomplish—the very same resources who are trying to complete the job in a timely manner. Therefore, grass-roots EACs can be accommodated only once or twice each year per program in order for such exercises not to interfere with the primary mission of completing the contractual effort. At the upper extreme, a grass-roots EAC may be done quarterly, but that frequency may well overtax the limited resources of any program, and have an adverse impact on successful contractual performance.

However, and this is the good news, the cost and schedule performance data generated by the earned-value C/SCSC activities provides an effective way of complementing the periodic (annual, semi-annual, or quarterly) "grass-roots" EACs done by the functional organizations. Without disrupting personnel in the performing organizations, and with the help of computer software programs in place today, a monthly (or even weekly) full range of EAC forecasts may be efficiently provided, based upon the actual C/SCSC performance data.

In addition, work being performed by all subcontractors is also subject to monthly EAC analysis, independent of what the supplier may be "officially" forecasting. Periodic subcontractor EACs should be verified independently by the responsible cost account manager (CAM), and one of the best ways to accomplish this is by examination of the supplier's earned-value performance. And should the CAM also be the same individual who approves all progress payment invoices, we will have established a critical linkage between the two management processes. The preparation of an independent EAC forecast for subcontractors also provides better assurance to the government that a "loss ratio adjustment" will be invoked at the appropriate time to preclude any overpayment of government funds.

In 1991, when the Department of Defense made their long-awaited changes to the C/SCSC requirements documentation and incorporated them directly into their new acquisition policy statement in DODD 5000.1 and DODI 5000.2, there were no changes to the C/SCS Criteria. There were, however, two important changes with respect to the analysis of the C/SCSC data, particularly as related to providing the estimates of costs at completion (EAC) on a given program.

The first change requires a "range of EAC estimates" to be provided by the military service program manager who is responsible for the management of a given acquisition system. The service proram manager must take the following steps based on the cost/schedule performance data:

1. Enter the range of estimates at completion, reflecting best and worst cases.[13]

The second DOD change to C/SCSC requires a justification, again from the military service program manager, whenever an estimate of costs at completion is forecasting a final value that is *less* than an amount using the cumulative Cost Performance Index (CPI) to forecast the EAC.

2. Provide the estimate at completion reflecting the best professional judgment of the servicing cost analysis organization. If the contract is at least 15 percent complete and the estimate is

lower than that calculated using the cumulative cost perform-ance index, provide an explanation.[14]

The "15 percent" thresholds referred to in the second quote re-lates to the DOD (Christle) empirical study of over 400 contractors who performed using C/SCSC, which was covered earlier. After performing 15 percent of a contract, what the contractor has achieved thus far is likely to be the *lower* end value of what they will do by the end of the contract.

With the added emphasis on using earned-value performance data to forecast the final cost/schedule outcomes of contracts, we would be wise to make sure that we fully understand some of the formulas available to forecast a "range of estimates." While there are amultitude of EAC formulas in use by the C/SCSC practitioners, we will address only the basic three, since these three constitute the more accepted methods in use.

We will discuss three EAC methods as follows:

1. The *low-end* Estimate at Completion (the mathematical EAC).

2A. The *middle-range* Estimate at Completion (the CPI EAC).

2B. What CPI performance factor it will take "to complete" an ef-fort, called the To Complete Performance Index (TCPI), in order to achieve the "CPI EAC" forecast.

3. The *high-range* Estimate at Completion (the CPI × SPI EAC).

We will address each of these mathematical forecasting meth-ods individually, building on the definitions of C/SCSC terms cov-ered earlier in this chapter.

Note: In order to follow this discussion it is necessary to un-derstand the C/SCSC jargon covered around the display of data in Figure 18.4, particularly the Cost Performance Index (CPI) and the Schedule Performance Index (SPI). A perfect CPI is 1.0, which means that for each dollar actually spent, one dollar of physical work was performed. A perfect SPI is also 1.0, which means that for each dollar of work planned to be accomplished, one dollar of physical work was accomplished.

If a contractor achieves what it sets out to achieve in its cost and schedule baseline plans (PMB), that is considered acceptable or even "perfect" efficiency in a performance measurement environment. This concept is illustrated in the diagram in Figure 18.7.

Figure 18.7. Monitoring Earned-Value (C/SCSC) Performance.

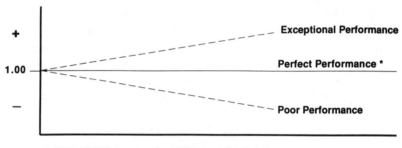

* Perfect Cost Performance: Cost Performance Index (CPI) = 1.0
$1.00 Cost Actuals = $1.00 Earned Value

* Perfect Schedule Performance: Schedule Performance Index (SPI) = 1.0
$1.00 Work Planned = $1.00 Earned Value

If after establishing the performance measurement baseline (PMB) the contractor achieves a cost performance factor of 1.0, this is considered excellent results. For every dollar they spend, they have achieved one dollar in physical earned-value accomplishments. Anything less than 1.0 is considered negative performance. Anything greater than 1.0 is considered positive or even exceptional efficiency.

On occasion, a contractor may actually achieve a performance greater than 1.0. It is sometimes possible for a contractor to perform slightly under 1.0 in the first part of a contractual period, then exceed 1.0 in the final stage. This may be the result of conservative planning of the performance baseline, where the final 100 percent achievement of various tasks is restrained in order to stimulate exemplary performance by program personnel.

However, if a contractor claims achievement "significantly" greater than 1.0, perhaps 1.5, then someone might want to pay the supplier a visit and find out how such "miracles" have occurred. Often, but not always, a performance attainment much greater than 1.0 is the result of an improper original performance measurement plan. And sometimes, exceptional performance is just plain "games-manship" by those who are preparing and/or approving the cost/schedule reports.

Likewise, schedule performance of 1.0 is considered as good as it can get, under normal circumstances. For every one dollar of work planned to be accomplished, one dollar of performance was achieved.

Now to address the range of EAC possibilities. The low-end EAC forecast is called the "Mathematical EAC," which is displayed in Figure 18.8. Some people refer to the Mathematical EAC formula as being "useless," or "unrealistic," and even "optimistic." And yet, many firms in the industry have been using this EAC method since

Figure 18.8. The "Mathematical" Estimate at Completion (EAC).

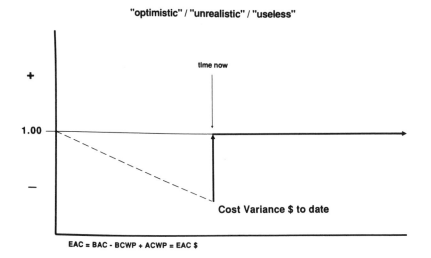

the C/SCSC were issued in 1967. The formula for the mathematical EAC is Budget at Completion (BAC), less the cumulative earned value (BCWP), plus the cumulative actual costs to date (ACWP). What this EAC does in effect is to "buy out" any poor performance to date, but assumes that starting tomorrow, all remaining work will be performed at perfect, or 1.0, efficiency (on the average).

While the mathematical EAC forecasting method is not a useful device for accurately forecasting what a program will likely cost at the end, it does provide a "floor" EAC, the value that represents the absolute minimum cost for the program. Such revelations sometimes come as a shock to management, and provide the lower-end range of EAC possibilities.

The middle-range EAC forecast is called the "Cumulative CPI EAC," which is displayed in Figure 18.9. The formula is the Budget at Completion (BAC), divided by the cumulative Cost Performance Index (CPI). There are a number of variations for this mid-range EAC, which will serve no value to us in this limited discussion of the subject. Some people will use only the last three or six months

Figure 18.9. The "Cumulative CPI" Estimate at Completion (EAC).

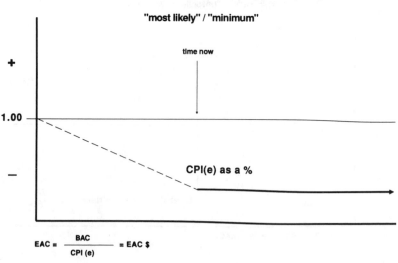

of the CPI, to reflect a recent trend or change the direction of the CPI. For our purposes we need only to understand that the total budget available is divided by the cumulative performance efficiency factor. If that efficiency factor is less that 1.0, then the estimate to complete the job will grow from the original allocated budget.

The CPI EAC is the most common and accepted EAC method. Some people consider this method to reflect the "most likely" EAC forecast, while other more conservative individuals feel it only reflects the "minimum" EAC. Nevertheless, in the recent DODD 5000.2-M, as quoted earlier, the military service program manager must now "provide an explanation" for any EAC forecasts that predict a final performance value that is less than that using the cumulative CPI EAC method.

One of the most important tools in C/SCSC forecasting does not deal with how much it will cost, or how long it will take to complete the job. Rather, the "To Complete Performance Index (TCPI)" has its utility in determining what performance efficiency factor it will take to do what you say you will do. Simply put, if you complete one-half of a job with a CPI of .95 percent, then one must assume that in order to complete the job within the approved budget for the remaining work, one must achieve a CPI of 1.05 percent for the balance of the effort. This concept is illustrated in Figure 18.10, as well as the formula to calculate the TCPI.

The value of the TCPI formula is that it can be used to answer a number of questions, all related to achieving some future objective. For example, what efficiency factor it will take: (1) to stay within the Budget at Completion (BAC); (2) to stay within the latest Estimate at Completion (EAC); (3) to stay within the latest Over the Target Budget (OTB); (4) to stay within the Fixed Price Incentive (FPI) ceiling, etc. The TCPI is used to "puncture" blind optimism, which sometimes inflicts our management, particularly our more senior management.

The high-end EAC forecast is called the CPI x SPI EAC forecast. It adds the dimension of scheduled but unfinished work, which was in the original plan, but has not been completed. This concept is illustrated in Figure 18.11, and the formula is the work remaining to be performed (BAC less BCWP), divided by the prod-

Figure 18.10. To Complete (the work) Performance Index (TCPI).

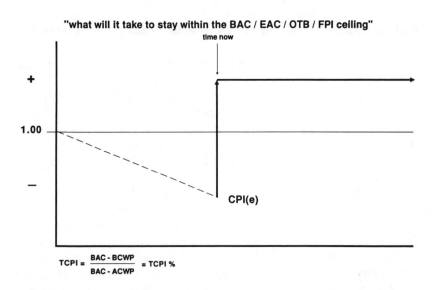

"what will it take to stay within the BAC / EAC / OTB / FPI ceiling"

$$TCPI = \frac{BAC - BCWP}{BAC - ACWP} = TCPI\%$$

uct of the Cost Performance Index (CPI) times the Schedule Performance Index (SPI), plus the ACWP. Obviously, if one performed under 1.0 in both the CPI and SPI, the resulting estimate at completion will be substantial.

This EAC method can get to be quite emotional to those involved in managing programs. Some consider this technique to represent the "most likely" EAC, while others call it the "worst-case" scenario. Some program managers, attempting to keep their costs under control refer to this technique irreverently as a "self-fulfilling prophecy."

This method is generally considered to be the high-end EAC method, and was used by the DOD cost analysts on the A-12 pro-

Figure 18.11. The "Cumulative CPI × SPI" Estimate at Completion (EAC).

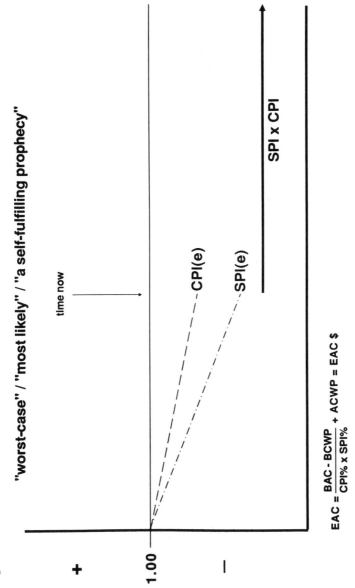

"worst-case" / "most likely" / "a self-fulfilling prophecy"

time now

+

1.00

CPI(e)

SPI(e)

SPI × CPI

−

$$EAC = \frac{BAC - BCWP}{CPI\% \times SPI\%} + ACWP = EAC\ \$$$

gram to forecast its total estimate at completion. It is a most valuable high-end EAC forecasting technique.

Now, what do all of these EAC methods do for us when we are attempting to forecast the final costs of a given program? If we take each of the formulas as displayed in Figures 18.8 to 18.11, and relate them to the data provided from the performance in Figure 18.4, perhaps we can discern the value in employing these EAC techniques.

Starting from the low-end EAC method we can line up a range of EAC forecasts.

1. The "Mathematical EAC":

$$BAC - BCWP + ACWP = EAC\$$$
$$\$100M - \$25M + \$40M = EAC\ \mathbf{\$115M.}$$

2A. The "CPI EAC":

$$BAC/CPI = EAC\$$$
$$\$100M/.625 = EAC\ \mathbf{\$160M.}$$

2B. The "TCPI":

$$(BAC - BCWP)/(BAC - ACWP) = TCPI$$
$$(\$100M - \$25M)/(\$100M - \$40M) = \$75M/\$60M = TCPI\ \mathbf{1.25.}$$

3. The "CPI × SPI EAC":

$$((BAC - BCWP)/(CPI \times SPI)) + ACWP$$
$$((\$100M - \$25M)/(.625 \times .500)) + \$40M = EAC\ \mathbf{\$280M.}$$

The low-end EAC method (mathematical EAC) tells us we will spend $115M, the minimum floor. The mid-range EAC (CPI EAC) forecasts a total cost requirement of $160M, and if anyone predicts a lesser amount, a military program manager will have to justify the lesser amount in order to comply with the new DOD 5000.2-M. The high-end EAC (CPI × SPI EAC) tells us we will need $280M to complete the job, quite an increase over our budget of $100M! With these formulas we can provide a full range of EAC estimates.

What is also significant is the fact that in order to stay within the original approved budget of $100M, the contractor must achieve a TCPI performance efficiency factor of 1.25 (or 125 percent) for all of the remaining effort. A very ambitious goal for any mortal person or group to achieve.

The various EAC forecasting methods available to us when earned-value methods are employed can be most beneficial in the management of our contracts or subcontracts, and provide a complementary adjunct in the effective administration of contractor progress payments.

In Summary

While there is no universal acceptance for the concept that there should be, or must be, a linkage between the progress payments being made to suppliers and the physical performance they are achieving on such work, the A-12 program experience will likely require such relationships in the future. This important linkage will likely improve both management processes: progress payments and cost/schedule performance measurement.

By linking physical performance measurement to the approval of progress payments, the prime contractor should avoid the potential cost risks of making overpayments to the supplier. Also, the FAR covering progress payments specifically require that there be some monitoring of the supplier's performance, in order to know when it might be necessary to adjust the repayment rate, or to suspend further progress payments, or to invoke the loss ratio, among other things.

By linking progress payments with C/SCSC performance measurement the buyer will have a better understanding of what the suppliers are actually achieving in satisfying their statement of work. To exclude progress payment data from the actuals being reported in the Cost Performance Report (CPR) does nothing but distort the desired earned-value measurements. And with the subcontracted portions (the buy content) becoming such a major part of most prime contractor dollars (upwards of 80 percent in some cases), the exclusion of progress payment data in performance measurement can reflect major distortions in the data being reported. A better approach, it is felt, would be to measure the physical performance of all suppliers that have progress payments, including firm fixed-price suppliers, and to incorporate these actual dollars into all cost performance reports (CPR).

We touched on a couple of issues rather hurriedly in the discussion above, and it might be beneficial for us to reemphasize these points. It is important that the buyer—the individual who has delegated procurement authority to issue the subcontract—be given the responsibility for the review and approval of each and every progress payment invoice before any such payments are made. This is a fundamental issue. Also important, it is believed, is the concept that this same buyer be held responsible for the total management of his or her subcontract, including functioning in the role of Cost Account Manager (CAM) for the performance measurement of the supplier.

If a prime contractor thus places the responsibility for (1) the full administration of progress payments, including the approval or disapproval of all payment invoices; and (2) the management of a subcontract earned-value cost account in a single individual, then the prime contractor will have achieved a "linkage" of both activities. This critical coupling should prevent any future overpayment of suppliers, in advance of their actual physical work accomplishments, and improve the overall subcontract management processes.

We hope we have made the case for employing this approach.

Endnotes

[1] Chester Paul Beach, Jr., Inquiry Officer, in a memorandum for the Secretary of the Navy, "A–12 Administrative Inquiry," November 28, 1990, p. 2.

[2] Ibid., pp. 3–4.

[3] Mr. Wayne Abba, Office of Acquisition Policy and Program Integration, Office of the Secretary of Defense, in public remarks made on the A–12 program, to the management systems subcommittee of the National Security Industrial Association, Costa Mesa, California, January 16, 1991.

[4] Mrs. Eleanor Spector, Director of Defense Procurement, Office of the Secretary of Defense, in Congressional testimony, April 9, 1991.

[5] *Los Angeles Times*, from Reuters, June 8, 1991.

[6] Robert R. Kemps, Humphreys & Associates, Inc., formerly with the Department of Defense and later the Department of Energy, in an article he wrote entitled "Solving the Baseline Dilemma," for the Performance Management Association newsletter, Autumn 1990.

[7] Russell D. Archibald and Richard L. Villoria, *Network-Based Management Systems (PERT/CPM)*, (New York: John Wiley & Sons, Inc., 1967), p. 475.

[8] Department of Defense Directive 5000.1, dated February 23, 1991, "Defense Acquisition"; Department of Defense Instruction 5000.2, same date, *"Defense Acquisition Management Policies and Procedures."*

[9] See Fleming, *Cost/Schedule Control Systems Criteria: The Management Guide to C/SCSC.* (Chicago: Probus, 1988).

[10] Gary E. Christle, Deputy Director for Cost Management, Office of the Under Secretary of Defense for Acquisitions, in a paper entitled "Contractor Performance Measurement—Projecting Estimates at Completion," Atlanta, Georgia, October 26, 1987. Data updated from 200 to 400 contracts from the Beach report, November 28, 1990, p. 6.

[11] Robert R. Kemps, Director of the Office of Project and Facilities Management for the Department of Energy (DOE), in his paper entitled "Cost/Schedule Control Systems Criteria (C/SCSC) for Contract Performance Measurement," at the Performance Management Association conference in San Diego, California, April 1989.

[12] William J. Hill, "Toward More Effective Management and Control of Contractor Payments," appearing in Defense Systems Management College's *Program Manager* magazine, January–February 1991, p. 21.

[13] DOD 5000.2–M, p. 16–H–6.

[14] Ibid., p. 16–H–6.

Section Seven

Managing Quality Improvement Projects

Chapter 19

Managing Quality-Improvement Projects

The Nature of Quality-Improvement Projects

Every effort at quality improvement in an organization can be called a project, since it is a one-time occurrence. This means, of course, that the tools and methods of project management are appropriate. It is also worth noting that managers of large projects must pay attention to the quality of work being done by the project team, with an eye toward constant improvement, so the methodology presented in this chapter can be applied to projects themselves.

As for quality-improvement efforts in organizations, many fail because of poor management of the project itself. This handbook presents the methods of managing projects which, if practiced, should reduce the probability of failure.

The second reason why quality-improvement efforts fail seems to be that they often cut across functions, as is true of most projects themselves, and we do not seem to know how to manage across functions.

Dan Dimancescu (1992) calls the divisions within organizations *chimneys*, an appropriate term. Each chimney is a sacred ground for its members and its chief. No one is supposed to tread on such hallowed ground. In the past, this has been called *empire building*, and we have known of the problems created, but we seem unable to overcome them.

Managing across functions is not the same thing as putting together a *multidisciplinary* task force with a leader who manages that group. Neither is it the same as the traditional matrix organization, which has for many years been considered almost synonymous with project management itself. Matrix creates a situation in which a worker has two bosses, always considered anathema for getting anything done. With cross-function management, the aim is to create *two bosses with one hat*, to use Dimancescu's term (1992, p. 148).

It is often this distinction that creates problems in matrix projects. Members from a number of chimneys are assigned to the project, but each still has a chimney manager. The chimney manager has his or her own organizational objectives to meet, besides supporting projects. Naturally, there is a tendency to give priority to one's own objectives, rather than to those of the project.

In cross-function management, each chimney manager must be committed to the same objectives—they must all wear the same "hat." That this is not easy to accomplish is an understatement at best. It is nearly impossible in many cases.

The problem boils down to a fundamental shortcoming of how an organization is viewed, together with one's role in it. Chimney managers are inclined to be unable to see outside their own chimneys (after all, the walls are pretty high). They do not see themselves as part of a larger *organism*, which cannot function properly unless all of its parts act in a *cooperative* manner.

In the U.S., we excuse this by claiming that we aren't good team players because we are *rugged individualists*. We say that this country was founded by such people, and that is true. When that premise is examined, however, we find that its conclusion is nonsense. The pioneers who settled this country were indeed rugged individuals, but when it was necessary to cross a river or build a house or barn, they pitched in and helped each other get the job done. In other words, they were good team players!

This is a highly significant point, and one that we must recognize and deal with if we are going to make cooperative efforts succeed. I would much rather have a team of rugged individualists than one composed of a bunch of *wimps!* But I want them to be cooperative, and the only reason we are not cooperative is that we let self-interest have a higher priority than group interest.

This is partly a function of the reward systems in organizations, and suggests immediately an obvious practice—if you want people to be good team players, you must reward them for cooperation, not for being rogues.

With that in mind, the remainder of this chapter will focus on the unique features of quality-improvement projects. So that all of the ideas will hang together, I will present the basics of quality-improvement processes. For experienced readers, you can skip this material, but it is presented for the general reader who may not have much exposure to the basic principles.

Quality-Improvement Concepts[1]

Processes and Systems

Work in organizations is conducted through *processes*. A series of tasks constitutes a process. Further, a string of processes forms a system.

Since the organization works through processes, you can improve your work only by improving processes. Better processes mean better quality, which translates into higher productivity.

If a quality-improvement team feels overwhelmed when it tries to analyze a process, it may be that they have focused on an entire system. The system can be broken down into processes, which can be studied. Generally speaking, a team should study only one process at a time.

Customers and Suppliers

All processes are designed to receive certain *inputs* from *suppliers* and act on those inputs in some way to provide an *output* to someone else who will be called a *customer*. Naturally, customers can be internal to the organization or the final, external customer. Using Dimancescu's term, the customer is whoever is *next in line*.

Definition of Quality

> The modern definition of quality is meeting the needs of customers.

Only the customer can really define what quality is to him or her, since quality is defined as meeting the needs of the customer. Engineering specs do not define quality, unless they lead to meeting the customer's needs.

Since the quality of one worker's output cannot exceed the quality of his inputs, you must build quality into every step of the process. Further, quality cannot be built into a product—it must be designed in! This means that quality always begins with the product or service developer. For this reason, Quality Function Deployment methods are necessary to arrive at an adequate understanding of what will constitute quality to the next-in-line.

There are two facets of quality that must be understood in order to achieve total quality performance. These are:

Quality of target values and features. Are you doing the right things? Are you delighting customers? Are they getting exactly the products and services they need, precisely when and how they need them? Only your customers can answer these questions.

> **Effectiveness is doing the right things.**

Quality of execution. Are you doing things right? How efficient are the processes employed to deliver your products or services?

Peter Drucker (1974) suggested that we must be sure that we do the right things first, then that we do things right. That is, focus on effectiveness first, then on efficiency, since it makes no sense to be 100% efficient if you are doing the wrong things.

> **Efficiency is doing things right.**

This is referred to in Total Quality Management (TQM) literature as *doing right things right*.

> We can do:
> right things right
> Right things wrong
> Wrong things wrong
> Wrong things right.

Employee Involvement and Teamwork

Every employee must be convinced of the need for quality and must actively participate in the improvement process. This is usually best accomplished through the use of cross-function quality-improvement project teams.

The need for teams is dictated by the fact that no single individual is likely to have the skills or knowledge to completely solve a process problem.

> **Juran says that a project is a problem scheduled for solution.**

Scientific Method

All quality-improvement methods employed today are based on the scientific approach or method. The central idea of scientific method is that actions are taken based on *factual data*, rather than on opinions, hunches, or preference. It also involves looking for root causes of problems, rather than trying to deal with symptoms, to seek permanent solutions rather than relying on quick fixes.

Complexity

Complexity is a general term for unnecessary work—steps that make a process more complex without adding value to the product or service. There are four kinds of complexity, as follows:

1. Mistakes/defects

Mistakes and defects result in rework or scrap. It has been estimated that as much as thirty percent of a manufacturer's total effort may be to do rework. That means that on the average one-third of the work force is re-doing what the other two-thirds do. One company put it this way: one of their factories could be said to spend all year turning out garbage.

In many general projects, the same thing happens. Because we have a tendency to adopt a ready-fire-aim to managing projects, we have to redo much of the work that was done wrong. This may mean that one of every three members of a project team is spending his or her time redoing what the other two people did originally!

2. Breakdowns/delays

Even when products and services are not harmed, work may be delayed because of breakdowns in machines or communications. These breakdowns reduce efficiency (productivity) and increase product or service costs.

3. Inefficiencies

Inefficiencies can result from the incorrect use of a tool or method, from poor work flow, and so on.

4. Variation

When products are highly variable, workers must add steps to deal with scrap and waste, or to rework salvageable materials. By reducing variation, quality is improved and costs are reduced. Most quality-improvement efforts eventually focus on reducing variation in processes.

Variation

Taguchi has illustrated the principles of variation with a bull's-eye example. Note the two bull's-eyes shown in Figure 19.1. Which marksman would you rather have on your team? Clearly, the marksman with the smallest spread is the better one. By simply adjusting the sights on her rifle a bit, she will place all of her shots in the center of the target. The marksman with the greater spread (variation), however, needs to work to achieve greater accuracy.

Figure 19.1. Bulls' Eye Illustrations.

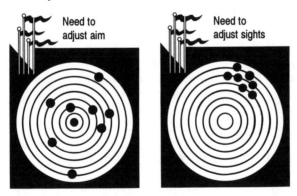

Common causes of variation. Common-cause variation is sometimes referred to as system noise. It usually results from a large number of small sources of variation. Common-cause variation determines the limits or capability of a process. All progress in engineering is an attempt to reduce random noise. In the chapter on estimating activity durations, we saw that it is noise that causes activity times to vary, and limits our ability to estimate with *precision*. Only by reducing the sources of that noise can our estimates be more precise.

Special causes of variation. Special causes are not part of the process all of the time. They arise because of specific circumstances. Failure to follow a standard procedure, for example, could lead to an increase in errors. In applying this concept to the improvement of quality within a project, sources of such variance can be outside vendor delays, improper work methods, and so on.

The two causes of variation must be handled differently. If you react to common-cause variation as if it were due to special causes, you will only make matters worse. Deming (1982) calls this *tampering*. If, however, you fail to notice a special cause, you will miss an opportunity to eliminate the source of problems.

Statistically Designed Experiments

The main feature of statistically designed experiments is that several factors can be tested simultaneously, rather than just testing one at a time, as in traditional experiments. By applying rules of probability, it can be determined which factor(s) under investigation account for the largest percentage of variance in a process, and thus which factors represent good candidates for attention in improving the process.

The Quality-Improvement Process

There are four steps in the quality-improvement process, usually called the Deming cycle, although Deming acknowledges Shewhart as the source. This cycle is called the PDCA cycle, the letters standing for Plan, Do, Check, Act. Juran (1989) has suggested that there are four steps in accomplishing the PDCA cycle, as follows:

1. Define the problem to be solved.

2. Diagnose the process to arrive at an understanding of the *root cause*.

3. Devise and implement a solution through problem-solving methods.

4. Continue monitoring the process to ensure that the problem remains solved.

Selecting a Quality-Improvement Project

Improvement projects are usually started because a manager or group of managers decides that change must occur. Projects that do not have top-management support are often doomed to failure before they get off the ground.

Projects should be selected only after widespread consultation with others in the company. If the project is to be a pilot, a lot is at stake, since a failure of the pilot is likely to convince people that the effort is a waste of time.

The following guidelines (adapted from Scholtes, 1991) should help in selecting initial projects, and should increase your chances for ensuring a successful project.

Select a process that

☞ Has a direct impact on the company's external customers.

☞ Repeats once a day or so. This will cause the effects of whatever changes you make to be visible within a few weeks.

☞ Is not already undergoing major revisions or is being studied using other methods (unless the project is aimed at understanding how to make a transition).

Common Errors in Selecting Projects:

➤ Selecting a process that no one is really interested in

➤ Selecting a desired solution, instead of a process

➤ Selecting a process already being modified

➤ Selecting a system to study, rather than a process

☞ Is relatively simple, with clearly defined starting and ending points. If you must target a large or complex system, break it down into smaller components.

☞ Is something a substantial group of managers agree is important to the company and its customers.

Select an area

☞ Where the managers, supervisors, and operators are likely to cooperate in the effort.

☞ That is highly visible in the organization, so the results will be noticed by people external to the area.

Balance projects so that

☞ At least half the projects have the potential to realize significant, measurable dollar savings.

☞ At least half, but not necessarily all, of the projects involve people who are relatively low in the organization's hierarchy.

☞ At least one is a collaboration between departments, perhaps involving research or marketing.

Choosing Players

Although a quality-improvement project will probably affect most members of an organization, there are four key roles that must receive special attention. These are:

Guidance Team. This group supports the project team's activities, secures resources for them, and clears a path in the organization. The team usually has three to six members who have diverse skills, a stake in the project, authority to make changes in the process under study, and who have clout and courage. These individuals are usually managers who oversee and support the activities of one or more project teams.

Before the project begins, the guidance team should

☞ Identify project goals.

☞ Prepare a mission statement.

☞ Determine needed resources.

☞ Select the team leader.

☞ Assign the quality advisor.

☞ Select the project team.

During the project the guidance team should

☞ Meet regularly with the project team.

☞ Develop and improve systems that allow team members to bring about change.

☞ When necessary, "run interference" for the project team.

☞ Ensure that changes made by the team are followed up.

Team Leader. This person runs the team, arranging logistical details, facilitating meetings, etc.

Quality Advisor. A person trained in the scientific approach and in working with groups. Helps keep the team on track and provides training when needed.

Project Team Members. Usually up to five members per project. The nature of the project dictates who they are. Usually they will be people who work closely with some aspect of the process under study. They often represent different stages of the process and groups likely to be affected by the project. They can be of various ranks, professions, trades, classifications, shifts, and work areas.

Preparing for a Project

The managers who select the project or the guidance team must consider some important issues before the project team can get underway.

They must:

Identify project goals. What is expected to be the result of this project? Avoid using numeric goals. But if they are used, make sure everyone knows they only indicate a magnitude of the desired improvement and must never be used as a measure of the team's performance.

Prepare a mission statement. The statement should tell the team:

☞ What process or problem to study.

☞ What boundaries or limitations there are, including limits on time or money.

☞ What magnitude of improvements they are asked to make.

> **Steps Needed to Prepare for a Project:**
>
> ➤ **Identify the goals**
>
> ➤ **Prepare a mission statement**
>
> ➤ **Determine needed resources**
>
> ➤ **Select the team leader**
>
> ➤ **Assign the quality advisor**
>
> ➤ **Select the project team**

☞ When they are scheduled to start the project and (if appropriate) the desired completion date.

☞ What authority they have to call in co-workers or outside experts, request equipment or information that is normally not accessible to them, and make changes in processes.

☞ Who is on the guidance team.

☞ How often they are expected to meet with the guidance team (should be about monthly for one or two hours) and the date of the first joint meeting (usually one month after the project team begins its work).

Determine the resources. What training is needed? Budget? Equipment? Which in-house or external specialists will be needed to advise the team? How much time must be allotted so team members will be able to complete the project? How will their normal work get done? By whom?

Select the team leader. The leader is often the manager responsible for the unit where most of the changes are likely to occur.

Assign the quality advisor. The guidance team should assign someone experienced in working with groups, and who knows how to teach others the basic scientific tools needed.

Select the project team. Ideally team members should represent each area affected by the improvements and each level of employee affected. Typically, teams should have no more than five members, in addition to the leader and quality advisor. Be cautious of having high-level managers on the team. Such individuals may intimidate members by their very presence.

> As a final guideline, don't expect team members to take on project work as additional work: adjust workloads to make time for the project.

Endnotes

[1.] Many of the ideas in this chapter are based on the work of Scholtes (1988). For an overall approach to quality improvement the reader is referred to his book, which can be ordered by calling 800-669-8326, or write Joiner Associates, P.O. Box 5445, Madison, WI 53705-0445.

Section Eight

Tools for Project Managers

Chapter 20

Project Scheduling Software

Sources of Software

If you are trying to locate a project scheduling program, the sheer number of available packages is a bit overwhelming. When I conceived this handbook, I had originally intended to create a comparison matrix so readers could save themselves the effort of creating one themselves. However, software goes through revisions so often that I realized that before I could get the handbook published my matrix would be obsolete. For that reason, I cover how to evaluate software in chapter 21, and provide a *fill-in-the-blank* matrix for your use in doing that evaluation.

What occurred to me was that it can be very difficult to locate information on software when you want it, so I decided to include the names of all of the software of which I am personally aware, together with the company name, address, and phone number, so readers can obtain information on those packages that interest them. The listing that follows is arranged alphabetically by software name. I have also listed *price codes,* as employed by some record stores, so you will have some idea of approximately the suggested list price for the software. Again, I have done this because changes in pricing will make an exact listing obsolete very soon, whereas I believe a price code will remain valid for a while longer.

Software publishers are welcome to send me information on their products for inclusion in future revisions of this handbook. My address is listed in the Preface.

Price Codes:

A = Under $200   B = $201–$600   C = $601–$1,200

D = $1,201–$2,000   E = Over $2,001

Product Name	Company & Address	Phone No.	PC/MAC	Code
Artemis Project	Metier 2900 North Loop W. #1300 Houston, TX 77092	713-956-7511	PC	E
Harvard Project Manager	Software Publishing Corp. 1901 Landings Drive Mountain View, CA 94039	415-962-8910	PC	C
InstaPlan 5000	Micro Planning International 655 Redwood Hwy. #311 Mill Valley, CA 94941	800-852-7526 FAX: 415-389-8046	PC	B
Micro Man II	POC-IT Management Services 429 Santa Monica Blvd. #460 Santa Monica, CA 90401	213-393-4552	PC	E
Micro Planner	Micro Planning International 655 Redwood Hwy. #311 Mill Valley, CA 94941	800-852-7526 FAX: 415-389-8046	PC & MAC	D

Product Name	Company & Address	Phone No.	PC/MAC	Code
Microsoft Project	Microsoft Corporation One Microsoft Way Redmond, WA 98052	206-882-8080	PC & MAC	C
Primavera	Primavera Systems 2 Bala Plaza Bala Cynwyd, PA 19004	800-423-0245 215-667-8600	PC, VAX	E
Project/2	Project Software & Development 14 Story Street Cambridge, MA 02138	617-661-1444	PC	D
Project Scheduler 4	Scitor Corp. 393 Vintage Park Dr. #140 Foster City, CA 94404	414-570-7700	PC	C
Project Workbench	Applied Business Technology 361 Broadway New York, NY 10013	212-219-8945 FAX: 212-219-3925	PC	D
Promis	Strategic Software Planning 150 Cambridgepark Drive Cambridge, MA 02140	800-783-1504 617-354-1504	PC	D
Quicknet Professional	Project Software Development 20 University Road Cambridge, MA 02138	800-366-7734 617-661-1444	PC	D

Product Name	Company & Address	Phone No.	PC/MAC	Code
SuperProject Expert	Computer Associates Int'l Inc. 1240 McKay Drive San Jose, CA 95131	800-531-5236 FAX: 408-432-0614	PC	C
Time Line	Symantec Corp. 10201 Torre Ave. Cupertino, CA 95014	800-441-7234 800-626-8847	PC	C
ViewPoint	Computer Aided Management 1318 Redwood Way Petaluma, CA 94954	800-635-5621 707-795-4100	PC	E

Chapter 21

Methodology for Evaluating Project Scheduling Software

Overview

Following the development of spreadsheets, word processing, and database software for personal computers, there has been an explosion in the development of project scheduling software. At one time, there were around 150 packages known to exist; as of this writing (October 1992) the number available is not known, but probably six packages account for 95% of all sales.

Still, there are a lot of packages available, and trying to select one that will meet your needs is like trying to navigate through a strange country with no road map. The proper approach is to set up a comparison matrix and enter data for the packages being considered and make a selection based on the unit that has the biggest "bang for the buck."

Some years ago, Applied Business Technology Corporation, the producers of Project Workbench™, developed a methodology for evaluating software, which was intended to be used by their clients to compare Workbench to other programs. Their methodology is reduced to a comparison matrix, into which you can place checkmarks or "yes-no" responses. However, the matrix is based on philosophy which seems sound, and with their permission, the following guidelines are extracted from their methodology.

The evaluation of any management productivity tool should be closely related to the management process that it impacts. For project management software, the software should be closely related to the role of the project manager—that is, to what the project manager actually does. The methodology is therefore driven by the "business of project management," rather than disconnected "features" that are unrelated and perhaps irrelevant to the day-to-day activities of a project manager.

This approach is different from that taken by many trade publications in which evaluations are totally feature-driven (e.g., "Does it have a relational database?" "Does it print the Gantt chart sideways?" and so on). The features approach is misleading, since it does not ensure that the feature is properly designed or even useful to a project manager for getting the job done.

The evaluation methodology should relate the functionality of the software to the following activities that are normally performed by a project manager:

☞ Creating an initial project plan

☞ Refining that plan

☞ Tracking status versus plan

☞ Replanning the remainder of the project

☞ Reporting status

☞ Supporting departmental management

☞ Using other tools/methodologies of the organization

The methodology developed by ABT is based on these seven functions. Each function is stated in the form of a question, and each of these is supported by a set of detailed sub-questions. "How easy is it to create the initial plan?" would be an example of a basic question directly related to the role of the project manager. "Can a project manager enter incomplete data about an activity?" is an example of a detailed question supporting the basic question shown above. A detailed question may support more than one basic question.

The complete methodology examines the following areas:

☞ **Summary:** summarizes the major findings about the product, its primary marketplace, what it does, how it does it, and its strengths and weaknesses.

☞ **Systems Architecture and User Interface:** describes the architecture of the system (i.e., its command structure and data-entry modes) and considers how easy it is to learn and use the system. Also included is an assessment of the user manual, tutorial, and help function.

☞ **Functionality:** considers the capabilities of the software, structured in the basic/detailed question format discussed above. The basic questions are as follows:

☞ How easily can the initial plan be created?

☞ How easily can the plan be refined?

☞ How easily can the remainder of the project be replanned?

☞ How well is information reported to the project team and to management?

☞ How well is department management supported?

☞ How well are other tools/methodologies interfaced to or made use of?

☞ **Capacity Limitations:** considers the limits in terms of the number of tasks, resources, etc., which can be handled by the system.

☞ **Marketing Information:** covers the pricing structure, support, training, user base, and general corporate information.

The matrix based on the ABT methodology contains one important feature: the first column asks the evaluator to decide if a feature is a "must" or "want." This allows the easy elimination of any package that does not provide a *must* feature.

Since the matrix that follows was developed by ABT so that users can compare Project Workbench™ with other packages, naturally the answer to each question is "yes" for PW, so there is no need to enter it into the matrix. This matrix allows seven packages, then, to be compared with each other and with Workbench. Also note that some of the features of Workbench are trademarked, such as the Applications Gateway.

Project Management Software
Evaluation Checklist

Ease of Use (Blank lines for user entries)	Must/Want Criteria					
"Toggle" fields to prevent incorrect data						
Context-sensitive help with "related topics" for browsing						
Clear, intuitive forms with "rolodex" feature for clarity						
Complete documentation with multilevel tutorial						
Automated Guided Tour						
Ability to reconfigure Gantt Chart as desired						
Blank lines permitted on Gantt Chart for clarity						
Screen colors under user control						
Planning						
Resource Spreadsheet™ on same screen as Gantt Chart						
CPM/PERT network created automatically						
Built-in project hierarchy (Work Breakdown Stucture)						
An organization's own breakdown structure can be used						
Gantt Chart shows % Complete, Actuals Thru, or % Work Expended						
Variety of timescales (days, weeks, quarters, years)						
Global, project, and individual resource calendars						
Variety of timescale options (start of week, days in week)						

Planning, continued	Must/Want Criteria					
Resource usage entered independently of task duration						
Up to 200 resources assigned to one task						
Resources can be assigned to multiple tasks concurrently						
Tasks can be organized in categories						
Up to 63 characters in the task field name						
Task duration calculated by system to shorten project and maximize resource utilization.						
Split tasks (start, stop, start) under user control						
Critical path calculation based on original or revised dates						
Dependency Definition Diagram						
Finish-start, start-start, & finish-finish dependencies						
Gap/Overlap (Lead/Lag) relationships						
Reconfigure CPM network manually						
Tasks can be independent of the CPM network						
Copy a task (or range of tasks) in project hierarchy						
Move or reorder a task (range of tasks) in project hiearchy						
Shift task & see effect on resource usage immediately						
Length/shorten tasks interactively						
Set schedule start date without locking a task in place						
Automatically create schedules with resource constraints						

Resource Management:	Must/Want Criteria					
Automatically level resources according to task priorities						
Set resource usage to maximum percentage of availability						
Insert a project, or part of a project, into another project						
"Go-to" feature for navigating through large projects						
All types of resources (staff, equipment, materials, cost)						
Fractional hours/days permitted for resource usage						
Flexible resource availability histogram						
Discontinuous tasks and work effort						
Organize resources into categories						
Assign resources with varying availabilities						
Assign resources to a task using any one of five loading patterns (front, back, uniform, fixed, contour)						
Automatically assign resources non-uniformly to task to absorb unused resources						
Summarize resources across multiple projects						
Display resource utilization, unused availability, or total availability on screen						
Analyze resource costs by period						
Tracking:						
Track against the original plan (baseline)						

Tracking, continued:	Must/Want Criteria						
Remove or reset baseline							
Track actual/estimated start and end dates							
Track percentage complete							
Track actual usage for all types of resources by period							
Track resources across multiple projects							
Use actuals to calculate project variances							
Actual resource usage independent of task duration							
Actual captured even if in excess of availability							
Actuals captured for unplanned tasks							
Variably load "estimate to complete" per resource per task							
Replan resource usage (keeping the original)							
Display and compare original & revised plans simultaneously							
Analysis & Reporting:							
Library of eight standard reports							
Library of 27 customizable reports included with system							
Full customizing capabilites for ad hoc queries							
New reports can be designed from scratch							
Columnar and tabular reporting options							

Analysis & Reporting, continued:	Must/Want Criteria					
Tabulate by data values, data ranges, or timescales						
Earned value and variance reporting						
Logical and nested data selection criteria						
Up to 11 data selection criteria permitted per report						
Up to 12 sort options permitted per report						
Forecast-based status reports						
Variance-to-baseline reports						
Display task status on CPM network						
Report scheduled vs. late task						
Annotate reports with text						
Preview reports online with full scroll/text editing capability						
Analyze cash-flow projections						
Turnaround time sheet						
Plot CPM/PERT Network						
PostScript™ printer support						
ASCII data export to other applications						

Multiple Projects:	Must/Want Criteria					
Multiple reports in user-defined levels of detail						
Multiproject resource spreadsheet						
Level resources across multiple projects						
Create a master project from subprojects or parts of subprojects						
Upload resource utilization profile from subproject o master project						
Automatically create "summary" resources in master project						
Establish interproject dependencies						
LAN support (including file locking & interstation messaging)						
Interfaces:						
Compatible with Windows 3.1 as a DOS application						
Wide variety of import/export options (fixed-format, DIF, DSV, PRN)						
Filter export data through Project Analyzer™						
Enhanced dBase support (Applications Gateway™)						
Link to Excelerator™ CASE tool						
Utility for converting Time Line™ project data into Project Workbench files						

Limits:	Must/Want Criteria						
Unlimited maximum number of tasks per project							
Up to $21 million maximum cost per task							
Up to 200 resources per task							
Up to 100 years total project length							
Up to 900 dependency links per project							
Up to 40 projects in one project group							
Support:							
The client support program provides:							
A hotline staffed with professionals							
All upgrades during the subscription period							
Full 12-month warranty period							
Software productivity tools, including Applications Gateway software and the Applications Guide							
Fax Mail services							
A quarterly newsletter							
Active user groups nationwide							

Support, continued:	Must/Want Criteria					
Extensive training and consulting services						
Rapid implementation programs						
Other Features Important to User:						

Chapter 22

Resources for Construction Project Managers

The R.S. Means Company offers a number of tools for construction project managers. Among these are estimating tables (commonly called *means tables*), training seminars, software, and consulting services. You can obtain their catalog by calling 800-448-8182 or writing them at P.O. Box 800, Kingston, MA 02364-0800. Following is a sampling of some of the books that they offer in their catalog:

Avoiding and Resolving Construction Claims
Bidding for the General Contractor
Building and Managing Government Construction

The Building Professional's Guide to Contract Documents
Business Management for the General Contractor
Construction Delays: Documenting Causes, Winning Claims,
    Recovering Costs
Construction Paperwork: An Efficient Management System
Contractor's Business Handbook
Cost Control in Building Design
Cost Effective Design/Build Construction
Estimating for the General Contractor
Fire Protection: Design Criteria, Options, Selection
From Concept to Bid: Successful Estimating Methods
Fundamentals of the Construction Process
Hazardous Material and Hazardous Waste: Construction Ref-
    erence Manual
Home Improvement Cost Guide
HVAC Systems Evaluation
HVAC: Design Criteria, Options, Selection
Insurance Repair: Opportunities, Procedures, and Methods
Means Structural Steel Estimating
Means Plumbing Estimating
Means Interior Estimating
Means Repair and Remodeling Estimating
Means Construction Cost Data
Means Forms for Building Construction Professionals
Means Scheduling Manual
Means Forms for Contractors
Means Legal Reference for Design & Construction
Means Unit Price Estimating
Means Square Foot Estimating
Means Landscaping Estimating
Means Mechanical Estimating
Means Labor Rates for the Construction Industry
Means Facilities Cost Data
Means Construction Cost Indexes
Means Assemblies Cost Data
Means Open Shop Building Cost Data
Means Repair and Remodeling Cost Data
Means Residential Cost Data

Means Mechanical Cost Data
Means Plumbing Cost Data
Means Concrete Cost Data
Means Landscape Cost Data
Means Site Work Cost Data
Means Electrical Estimating
Means Interior Cost Data
Means Graphic Construction Standards
Means Electrical Change Order Cost Data
Means Illustrated Construction Dictionary
Means Man-Hour Standards for Construction
Means Square Foot Costs
Means Light Commercial Cost Data
Means Estimating Handbook
Project Planning and Control for Construction
Quantity Takeoff for the General Contractor
Risk Management for Building Professionals
Roofing: Design Criteria, Options, Selection
Superintending for the General Contractor
Survival in the Construction Business: Checklists for Success
Understanding Building Automation Systems
Understanding the Legal Aspects of Design/Build

Chapter 23

Associations for Project Managers

Following is a listing of some of the professional associations that may be of interest to project managers. Information is provided on how to contact them, with no endorsement being made by me as to whether they are worthwhile.

Project Management Institute
P.O. Box 43
Drexel Hill, PA 19026-3190
215-622-1796
FAX: 215-622-5640

PMI has local chapters throughout the United States. Contact the main number listed above for information on whom to contact in your area if you are interested in a chapter.

Internet
Secretariat
Internet/CRB Switzerland
Zentralstrasse 153
Zurich CH 8003 Switzerland

American Management Association
135 West 50th Street
New York, NY 10020

Engineering Management Society of
Institute of Electrical & Electronic Engineers
345 E. 47th Street
New York, NY 10017-2366
212-705-7900

National Management Association
2210 Arbor Blvd.
Dayton, OH 45439-1580
513-294-0421

American Society for Training & Development
1630 Duke Street
Alexandria, VA 22313
703-683-8100

Chapter 24

Checklists for Managing Projects

Project Planning

1. A problem statement has been written for the project.

2. The project mission has been communicated to all participants.

3. Risks have been identified and contingencies developed where possible.

4. Project strategy has been tested for P, C, T, S feasibility.

5. Force-field analysis is satisfactory.

6. Consequences have been analyzed and are acceptable.

7. The ultimate purpose of the project is understood by all team members.

8. At least one of the P, C, T, S variables is estimated, rather than all four being dictated.

9. Clear definition(s) of project performance requirements exist.

10. Adequate criteria exist for measuring achievement of performance targets.

11. Work Breakdown Structure has been developed to levels sufficient to permit estimates of cost, time, and resource requirements at desired accuracy.

12. WBS has been reviewed with client, contributors, senior management.

13. Schedule milestones have been established with planned reviews.

14. Task-level schedule has been developed against the WBS in network form.

15. The critical path has been identified.

16. The critical path allows the required end date to be met.

17. The critical path has been examined to determine if it is realistic.

18. A Gantt chart has been developed to be used as a working tool.

19. Resource allocation has been checked to ensure that no one is overloaded.

20. Resources are not allocated at more than 80 percent productivity.

21. Resource conflicts with other projects have been eliminated or resolved.

22. The control system has been designed.

23. Measures of progress have been established.

24. People who must implement the project plan participated in preparing it.

25. The plan is at the right level of detail (neither too much nor too little).

26. Estimates are based on recorded data for similar tasks when possible.

27. Padding of estimates has been done above-board.

28. Padding is acceptable to management.

29. Project plan has been reviewed in a signoff meeting.

30. Project notebook has been signed off by all stakeholders.

31. Concerns raised in the signoff meeting have been addressed to satisfaction of everyone.

32. The plan contains the following:

    ☞ problem statement

    ☞ mission statement

    ☞ project strategy

    ☞ project objectives

    ☞ QFD analysis or other means of identifying customer needs

    ☞ SWOT analysis

    ☞ statement of project scope

    ☞ list of deliverables and other contractual requirements

    ☞ end-item specifications to be met

    ☞ a work breakdown structure

    ☞ both milestone and task-level schedules

☞ resource requirements

☞ control system, including change control procedures

☞ major contributors in Linear Responsibility Chart form

☞ risk analysis with contingencies

☞ statements of work as required

33. Resource allocations include deductions for vacations, holidays, sick leave, etc.

34. Cost estimates include travel and living expenses if they will be required.

35. Costs for project security are included if appropriate.

36. Plans include time for reviews, meetings, approvals, and so on.

37. All physical facilities are expected to be available.

38. Testing facilities are adequate.

39. If new hires will be required, steps have been taken to ensure availability.

40. All project team members are qualified for their work.

41. Any required training of team members has been budgeted and provided for.

42. Any potential political problems that might affect this project have been identified and can be handled.

43. Arrangements have been made to promote free and open communication between all members of the team.

44. Where necessary, members have been co-located to facilitate communication. Where physical co-location is not possible, *virtual* co-location has been arranged.

45. Vendors have been required to submit their own project plans to ensure that all deliveries can be met.

46. Boundaries have been pre-established for change control.

47. System has provided that all project revisions will be distributed to all appropriate individuals/departments/parties.

48. Chart-of-account numbers have been set up for all project work.

49. Schedule and chart-of-accounts are traceable to the Work Breakdown Structure.

50. Unbudgeted project expenditures must be approved by the Project Manager.

51. Functional managers must inform the PM before reassigning personnel to other jobs.

52. Critical ratios have been established to aid project monitoring.

53. System in place to revise the project budget both upward and downward when appropriate.

54. All team members have personal plans for conducting their part of the project work.

55. Variance limits have been established for all contributors.

56. Bonus/penalty arrangements have been applied to vendors if needed.

57. A vendor certification program is followed to ensure vendor capability.

58. Critical future events have been evaluated for project impact.

59. Resource usage has been smoothed as much as possible.

60. The initial plan does not require significant overtime to meet initial schedule dates.

61. All deliverables have been identified for each milestone (including reports, etc.)

62. Performance specs have been written and agreed upon by all stakeholders.

63. Government regulations (and others) have been identified and cited in the project plan.

64. For product design projects, representatives from manufacturing are part of the team.

65. The *real* customer has been consulted in order to pin down requirements.

66. SWOT analysis is based on data, rather than strictly personal biases, etc.

67. Team members have been selected whose individual needs will be met through participation in the project (where possible).

68. Project termination procedure has been developed.

69. Team members have been convinced of the value of the project goals.

70. Controls are not so rigid that they stifle innovation.

71. Previous project records for similar programs have been reviewed before planning this one.

72. Unique physical resources (such as test equipment) have been entered into the schedule so bottlenecks can be spotted.

73. Required resources that do not yet exist have been identified as risks to project success.

74. Roles and responsibilities of each team member have been clearly defined.

75. Procedures for doing work have been developed by participants and approved by managers.

76. Performance specifications greater-than-required have not been asked for.

77. Tasks with durations greater than 4-6 weeks have been subdivided to avoid back-end loading.

78. Parallel critical paths have been eliminated if possible.

79. Network diagrams have been checked for logic violations.

80. Functional managers in matrix projects have resource loading diagrams to support their ability to staff projects.

81. Projects that span long periods have been budgeted to account for inflation.

82. Exit criteria have been established to define completion of each project phase.

Execution Checklist

1. Meetings are scheduled on a regular basis to review progress.

2. An auditor has been assigned for all audits.

3. Estimates of progress on nonquantifiable work are checked by independent party.

4. Estimates of work remaining are not just linear projections—unless those can be justified.

5. Causes of delays and other problems have been explained in progress reports and documented in the project notebook.

6. Impact of scope changes has been explained to stakeholders and approved.

7. Impact of unexpected resource shortages has been computed and explained.

8. All team members have been trained in the use of Earned-Value Analysis.

9. Meetings have been scheduled for the project team to look at improving their work processes.

10. Transfer or termination of a team member has been coordinated with his or her replacement.

11. For cases where coordination is impossible, the predecessor has left written instructions for his or her successor.

12. Progress reports show "red flags" for situations expected to have serious impact on project performance.

13. Monitoring of outside vendors is periodic.

14. Progress payments to vendors are based on earned-value analysis.

15. Team reviews are facilitated by an independent person.

16. Action plans are in place to address the outcomes of team review meetings.

17. Action assignments have been made to follow up team meetings.

18. Time worked is logged daily by contributors to the project.

19. All hours worked on a project are tracked back to the project, including nonpaid overtime hours.

20. Competition is kept to a minimum within the project team.

21. Where concurrent engineering or concurrent project management is applied, frequent coordination meetings are held to keep everyone informed.

22. Corrective action for off-target tasks has been developed and approved.

23. Progress reports are distributed in appropriate increments.

24. The expression, "If it ain't broke, don't fix it," is repeatedly challenged.

25. There are no penalties for performance that is better than the plan.

26. A climate of open discussion and inquiry exists in the project team.

27. Team members are encouraged to provide "early warnings" about developing problems.

28. Project Manager (PM) keeps *all* team members as fully informed as possible.

29. Decisions are made at the lowest possible level in the project.

30. Consensus decisions are made when appropriate, but not *every* time a decision is required.

31. A structured problem-solving approach is employed.

32. Taguchi methods are applied in design projects.

33. Functional managers are kept informed of changes that may impact them.

34. If the project is a disaster, PM has a current resume ready.

35. Where no contingency exists for a risky task, precautions are taken to minimize the risk.

36. Critical path activities are being managed so that they complete *at least* on time, and earlier if possible.

37. Tasks with float are completed at earliest times when possible. Float is reserved to handle unforeseen problems.

38. Memos to team members require RSVP to ensure that they were received.

39. Deliverables are used as milestone measures.

40. Actual project costs compare well with planned costs.

41. Personnel problems such as absenteeism, turnover, etc., are being addressed in a positive way, rather than being ignored.

42. Decisions made by the project manager are being accepted without complaint.

43. Morale in the team seems to be good.

44. Change procedures are being followed.

45. The customer(s) is involved and aware of project status.

46. Upper management is aware of project status.

Software Quality Checklist

1. All next-in-line parties have been involved in planning the project to ensure that their needs will be met.

2. Where appropriate, privacy, security, and audit matters have been taken into account in the design.

3. Adequate plans have been developed to ensure the architecture of the system is correct.

4. Testing will be done by an independent test group for objectivity.

5. Technical standards for design, coding, etc., are being followed.

6. The documentation is complete, understandable, and accurate, as certified by an independent auditor.

7. Primary deliverables are of satisfactory quality.

8. Deliverables meet customer requirements, as certified by the customer.

Project Change Procedure

1. There is a documented change procedure for the project.

2. Provision is made for handling requests for clarification and interpretation of existing documents.

3. Change requests are approved by the appropriate parties, with complete visibility by the project manager.

4. All change requests are evaluated for project impact and stakeholders are informed of impact before change is approved.

5. Resources allocated to the project are changed as necessary to accommodate project changes.

6. All changes are documented and stored in the project notebook.

Software Installation and Conversion

1. Conversions are audited by independent party to ensure quality.

2. An adequate recovery system exists in case data is lost during conversion.

3. Plans exist for maintenance of the new system.

4. Installation has been planned to have minimum impact on users.

5. New equipment and supplies required for the conversion have been identified and ordered.

6. Installation plans have been developed for new equipment.

7. Training program has been developed to ensure user capability with the new system.

8. There is a fallback plan in the event of conversion problems.

9. Arrangements have been made with outside providers of communications, etc., to ensure an on-time conversion.

Working Conditions Checklist

1. Adequate work space has been provided for all team members.

2. Lighting, temperature control, noise level, privacy, and safety issues have been addressed.

3. Adequate storage space exists.

4. A conference facility exists for team meetings.

5. Clerical support has been provided at adequate levels.

6. Provision has been made to stock adequate supplies.

Section Nine
Problem-Solving in Projects

Chapter 25

Problem Solving in Projects

Fundamental Concepts of Problem Solving

Definitions

> A *decision* is a choice of alternatives.

> A *problem* is a gap between where one is and where one would like to be, which is confronted by some obstacle that impedes movement to reduce the discrepency or gap.

A problem is a gap between where one is and where one would like to be!

Success in problem-solving depends upon the locating of obstacles that can be overcome.

Problems are often stated in terms of a desired goal. Lack of a desired goal is not a problem. An obstacle to attaining a desired goal constitutes a problem. Searching for ways to get around the obstacle constitutes problem-solving. The proper statement of a problem is one that stimulates searching behavior.

Creativity consists of shifting one's thinking from one obstacle to another (which can be overcome) rather than staying focused on finding a way to overcome a single obstacle.

Open- and Close-Ended Problems

> A close-ended problem is one that has a unique solution.
>
> An open-ended problem is one that may have a number of solutions.

An open-ended problem is one that has no single correct answer, and that has boundaries that can be challenged. A close-ended problem, on the other hand, has a single right answer. Most close-ended problems are typified by math problems or situations in which something once worked and has quit.

Our educational system teaches us to find the *one right answer* to problems presented, implying, perhaps, that most problems are close-ended. Actually, it is very likely that *most* of the problems we encounter in life are open-ended, yet our educational bias causes many of us to be unwilling or reluctant to challenge boundaries of real-world problems. Roger von Oech suggests that one way out of this dilemma is to always insist on finding the *second right answer!*

Problems encountered in projects may involve both open- and close-ended types. For example, if the process once worked but has become dysfunctional, you are dealing with a close-ended problem, since finding the cause of the dysfunction and fixing it involves a single solution. On the other hand, if the process is functioning correctly, but is to be improved, this may involve open-ended problem-solving.

Figure 25.1 should help you determine whether you are dealing with an open or close-ended problem.

In addition to the flowchart in Figure 25.1, the following table details the differences between close- and open-ended problems.

Open-Ended Problems	Close-Ended Problems
Boundaries may change during problem-solving.	Boundaries are fixed.
Problem-solving often involves production of novel and unexpected ideas.	The process has a predictable final solution.
May involve creative thinking of an unpredictable kind.	Process is usually conscious, controllable, and logically reconstructible.
Solutions are often outside the bounds of logic—can neither be proved nor disproved.	Solutions are often provable and can be shown to be logically correct.
Direct, conscious efforts at stimulation of creative process may be difficult.	Procedures are known that directly aid in problem-solving.

Figure 25.1. Recognizing open-ended problems (adapted from Richards; see reading list).

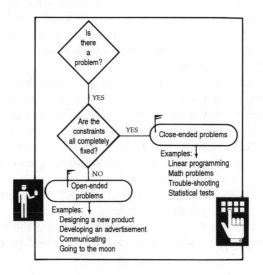

Chapter 26

Solving Close-Ended Problems

Using the Scientific Method to Define Problems

> The way a problem is defined determines its solution possibilities.

Before a problem can be solved, it must be defined. This seems clear enough, yet the educational system in the United States inadvertently produces in its students the tendency to be *solution-minded*, rather than *problem-minded*. Throughout their schooling, students are given problems to solve, and the teacher will generally accept only one right answer.

Then, when they leave school, people find that no one gives them the definition of the problem, and in many cases there is more than one right answer. For that reason, Americans tend to have difficulty with defining problems, and sometimes make the mistake of finding the right solution to the wrong problem.

As was stated in Chapter 3, the first major step in managing a project is to define the problem to be solved by the project. Since most projects are solving open-ended problems, the material in this chapter will be of limited usefulness at that step in managing a project. However, during implementation of the project, there will be many opportunities to apply the methods of solving close-ended problems.

Chapter 3 gave one example that illustrates the importance of defining problems, the case of the warehouse distribution system. As a second illustration, I will relate an experience of my own.

I was having breakfast in a hotel one morning, and overheard two men talking at the table next to me. It soon became clear that one of them was a district sales manager for a large corporation and the other was one of his young salesmen. The sales manager was clearly unhappy with his staff, and was giving his salesman a lecture. It went like this:

"The company has spent a great deal of money developing product x," he said, "and none of you are selling it. If you guys don't get busy and start selling the product, I'm going to get myself some salesmen who can sell!"

Well, it is pretty obvious how he has defined the problem, isn't it? He has himself a group of poor salesmen. So if they don't get busy and start selling, he is going to get rid of them and get some who can sell.

Now I don't know about anyone else, but I don't think his problem is with his salespeople. After all, how can he have *all* bad ones? Doesn't it seem reasonable that he should have hired at least one good one—even by accident?

But he claims to have all bad ones.

Let's give him the benefit of the doubt for a moment and assume that he is correct, and that he has indeed hired all bad ones. Suppose he gets rid of all of them and hires new ones. What do you suppose he will have?

"I'm going to get myself some salesmen who can sell!"

You bet! I'll bet he has all bad ones again.

But the fact is, I suspect there is something wrong with the product, or the market has changed, or it is priced wrong. If *none* of his salesmen can sell the product, it isn't likely to be the product. Nevertheless, he has defined the problem as people, and so the only solution open to him is to deal with the people.

This situation is more common than might be imagined. I frequently encounter people who have made up their minds what the problem is and are going about solving it, without having done a proper problem analysis to be sure that their definition is correct. For close-ended problems, the best approach to defining the problem is to use what is commonly called the scientific method, which consists of the following steps:

☞ Ask questions

☞ Develop a plan of inquiry

☞ Formulate hypotheses

☞ Gather data to test those hypotheses

☞ Draw conclusions from hypothesis testing

☞ Test the conclusions

Constructing a Good Problem Statement

1. The problem statement should reflect shared values and a clear purpose.

2. The problem statement should not mention either causes or remedies.

3. The problem statement should define problems and processes of manageable size.

4. The problem statement should, if possible, mention measurable characteristics.

5. The problem statement should be refined (if appropriate) as knowledge is gained.

Solving Close-Ended Problems with Problem Analysis

As was previously stated, close-ended problems have single solutions. Something used to work and is now broken. The remedy is to determine what has broken and repair it—a single solution. To solve close-ended problems, we use a general approach called *problem analysis*, which is presented in the following section of this chapter.

Conducting a Problem Analysis

The steps in the problem analysis process are shown by the diagram in Figure 26.1.

Identification

The first step in the problem analysis process is identification: "How do I know I have a problem?" As was previously stated, a problem is a gap between a desired state and a present state, which

Figure 26.1. Problem Analysis Steps.

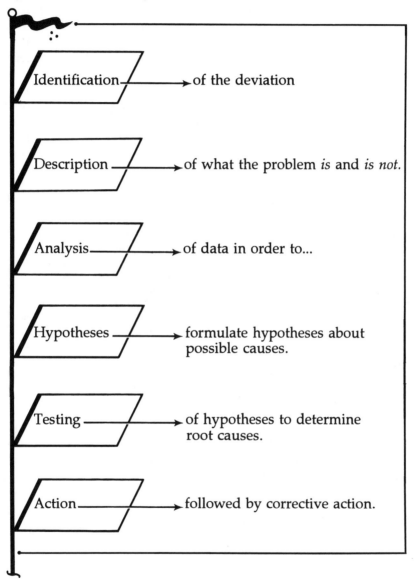

is confronted by an obstacle that prevents easy closure of the gap. That gap can be a *deviation* from standard performance when a process is involved. In monitoring progress in the project, when the critical ratio falls outside acceptable limits, this is a signal that a potential problem exists with the task in question. This is where problem analysis begins in that situation.

When dealing with deviations, we have to know the *norm*. How is the process supposed to behave? Some project work will have a great deal more variability than other project work. For that reason, critical ratio limits for some tasks might be set tighter than others. Once the normal variability is known, then we can determine if the deviation is significant, whether it is positive (performance is better than the norm) or negative (performance is worse than the norm).

In a more general sense, a problem is generally recognized because of the *effects* produced being different than the normal outcomes expected from the process. Those effects might be a change in scrap level, higher or lower production, or a drop in customer purchases.

In order to correct for the deviation, we need to find its *cause*. For a desirable deviation, we must know the cause so it can be replicated. For undesirable deviations, the cause must be remedied.

To determine the cause of the deviation, we employ a process called *description* of the problem.

> **Making Glass**
>
> Making glass is a nifty process. You melt sand to turn it into glass. Unfortunately, the process seems to be somewhat unreliable. Yields are often poor. A quality consultant went into a plant that had just achieved a yield of 85 percent one day, and the people were ecstatic! "How did you do it?" he asked, "We don't know." "Then how are you going to replicate it?" he asked.

Description Using Is/Is-Not Analysis and Stratification

Stratification and is/is-not analysis are ways to localize a problem by exposing underlying patterns. This analysis is done both before collecting data (so the team will

> **Stratum: a layer**

know what kind of differences to look for) and after collecting data (so the team can determine which factors actually represent the root cause).

To stratify data, examine the process to see what characteristics could lead to biases in the data. For example, could different shifts account for differences in the results? Are mistakes made by new employees very different f.rom those made by experienced individuals? Does output from one machine have fewer defects than that from another?

Begin by making a list of the characteristics that could cause differences in results (use brainstorming here). Make data collection forms that incorporate those factors, and collect the data. Next look for patterns related to time or sequence. Then check for systematic differences between days of the week, shifts, operators, and so on.

The is/is-not matrix on the next page is a structured form of stratification. It is based on the ideas of Charles Kepner and Benjamin Tregoe.

Analysis

Once stratified data have been collected, the differences can be analyzed so that hypotheses can be formulated as to causes of the problem. The following questions are designed to help identify differences:

☞ What is different, distinctive, or unique between what the problem is and what it is not?

☞ What is different, distinctive, or unique between where the problem is and where it is not?

☞ What is different, distinctive, or unique between when the problem is seen and when it is not?

The focus of these questions is to help us determine what has changed about the process. If nothing had changed, there would be no problem. Our search should be limited to changes within the differences identified above. The following question is intended to help keep us focused:

Figure 26.2. The Is/Is-Not Matrix.

	Is Where, when, to what extent or regarding whom does this situation occur?	**Is Not** Where does this situation NOT occur, though it reasonably might have?	**Therefore** What might explain the pattern of occurrence and nonoccurrence?
Where The physical or geographical location of the event or situation. Where it occurs or is noticed			
When The hour/time of day/day of week, month/time of year of the event or situation. Its relationship (before, during, after) to other events.			
What Kind or How Much The type or category of event or situation. The extent, degree, dimensions, or duration of occurrence.			
Who What relationships do various individuals or groups have to the situation/event. To whom, by whom, near whom, etc., does this occur? (Do not use these questions to place blame.)			

Instructions: Identify the problem to be analyzed. Use this matrix to organize your knowledge and information. The answers will assist you in pinpointing the occurrence of the problem and in verifying conclusions or suspicions.

What has changed about each of these differences?

Noting the date of each change may also help us relate the start of the problem to some specific change that was made to the process.

Hypotheses

At the heart of the scientific method is the testing of hypotheses based on the foregoing steps of data collection and analysis. A hypothesis is a guess or conjecture about the cause of the problem. At this point *all* reasonable hypotheses should be listed.

One of the most commonly used tool for formulating hypotheses is the Ishikawa or cause-effect diagram. It can be used separately or in conjunction with Is/Is-not analysis to help formulate hypotheses. The group technique employed will usually be brainstorming.

Test Hypotheses

To test hypotheses, we first ask if the suspected cause can explain both sides of the description. That is, the cause must explain both the *is* and the *is-not* effects. If it cannot explain both, it is unlikely to be a real cause. It may be possible to modify the assumption of "only if" to the statement.

The testing method follows:

☞ Test each possible cause through the description, especially the sharp contrast areas.

☞ Note all "only-if" assumptions.

The most likely cause will be the one that best explains the description or the one with the fewest assumptions. To be certain, you must now verify the hypothesis quickly and cheaply.

One test is whether you can make the effects come and go by manipulating the factor that is supposedly causing the deviation. If you can, you have probably found the true root cause.

Action

At this point, there are three possible actions that might be taken. These are:

☞ Interim action: To buy time while the root cause of the problem is sought. This action is only a Band-Aid™ for correcting symptoms.

☞ Adaptive action: You decide to live with the problem or adapt yourself to the problem.

☞ Corrective action: This is the only action that will truly solve the problem. It is aimed at the actual cause of the problem.

Chapter 27

Solving Open-Ended Problems

Problem Solving Through Creative Analysis

In solving project problems, it may be necessary to employ creative techniques to develop definitions, ideas, and so on. In particular, the problem being solved by the project itself is likely to be open-ended, requiring different methods than those presented in Chapter 26 for solving close-ended problems. Even the approach used to define the problem is different. For close-ended problems, the scientific approach to analyzing data can be used. For open-ended problems, however, we need different methods. The techniques presented in this chapter are intended to help problem solvers develop good definitions for open-ended problems and also to apply idea-generating aids which have been found useful.

I should mention here that Dr. Edward de Bono is considered by many people to be one of the leading experts on creative problem-solving, and his recent book, *Serious Creativity* (1992), covers the subject in more detail than this chapter can possibly do. I heartily recommend that the interested reader consult Dr. De Bono's works.

Redefinitional Procedures

The procedure outlined in Table 27.1 is designed to help you develop a good definition for an open-ended problem. However, it is only one approach, and others are presented following the table.

Table 27.1. An Exercise to Develop a Good Problem Definition.

1. Write down an open-ended problem that is important to you and for which you would like some answers that could lead to action. Take as long as you wish for this.

2 Again, taking your time, complete the following statements about the problem you have chosen. If you cannot think of anything to write for a particular statement, move on th the next one.

    a. There is usually more than one way of looking at problems. You could also define this one as . . .

    b. . . . but the main point of the problem is . . .

    c. What I would really like to do is . . .

    d. If I could break all laws of reality (physical, social, etc.), I would try to solve it by . . .

    e. The problem, put another way, could be likened to . . .

    f. Another, even stranger, way of looking at it might be . . .

3. Now return to your original definition (step 1). Write down whether any of the redefinitions have helped you see the problem in a different way.

The Goal Orientation Technique

Goal orientation is an attitude, first of all, and secondly, it is a technique to encourage that attitude. Open-ended problems are situations where the boundaries are unclear, but in which there may be fairly well-defined needs and obstacles to progress.

The goal-oriented person tries to recognize the desired end-state ("what I want") and obstacles ("what's stopping me from getting the result I want").

To illustrate the goal-orientation technique, consider the problem outlined in Table 27.2.

Table 27.2. Use of the Goal-Orientation Technique.

Original Problem Statement

Adult illitracy has reached alarming proportions. Ford Motor Company recently said they are having to train almost 25 percent of their work force in basic reading, writing, and arithmetic, at considerable cost.

Redefinitions:

1. (How to) efficiently and effectively teach adults to read.

2. (How to) keep kids from getting through school without being able to read.

3. (How to) get parents to take an interest in their kids so they will learn to read in school.

4. (How to) eliminate the influences that cause kids to take no interest in school.

The Successive Abstractions Technique

Suppose a company that makes lawn mowers is looking for new business ideas. Their first definition of their problem is to "develop a new lawn mower." A higher level of abstraction would be to define the problem as "develop new grass-cutting machines." An even higher level of abstraction yields "get rid of unwanted grass."

Another definition of the problem, of course, might be to "develop grass that grows to a height of only x inches above the ground." (See Table 27.3.)

Table 27.3. Successive Abstractions.

Highest level	Get rid of unwanted grass
Int. level	Develop new grass-cutting machines
Lower level	Develop new lawn mower

Analogy and Metaphor Procedures

One of the really interesting ways of describing problems is through the use of analogy or metaphor. Such definitions help increase the chances of finding creative solutions to problems. Such methods are especially useful in group techniques, such as brainstorming. In fact, they are actually preferable to literal statements, since they tend to be extremely effective in stimulating creative thinking. For example:

"How to improve the efficiency of a factory" is a down-to-earth statement.

"How to make a factory run as smoothly as a well-oiled machine" is an analogical redefinition.

"How to reduce organizational friction or viscosity" is a metaphoric definition.

Wishful Thinking

Many left-brained, rational people do not appreciate the value of wishful thinking. However, wishful thinking can provide a rich source of new ideas. Dr. Edward de Bono, in his work on lateral thinking, talks about an "intermediate impossible"—a concept that can be used as a stepping-stone between conventional thinking and realistic new insights. Wishful thinking is a great device for producing such "intermediate impossibles."

Rickards cites an example of a food technologist working on new methods of preparing artificial protein. As a fantasy, she considers the problem to be "how to build an artificial cow." Although the metaphor is wishful, it suggests that she might look closely at biological systems and perhaps look for a way of converting cellulose into protein, which is what takes place in nature.

Remember the statement from Table 27.1, "What I would **really** like to do is . . ." Or try, "If I could break all constraints, I would . . ."

Nonlogical Stimuli

One good way of generating ideas is through forced comparisons. This method can be used for developing ideas for solving a problem or as an aid to redefinition. In Table 27.4 is an example of the procedure, used in conjunction with a dictionary.

Boundary Examination Technique

When a problem is defined, the statement establishes boundaries as one sees them. If it is accepted that these are open to modification, then the definition is only a starting point. Unfortunately, many people do tend to treat boundaries as unchangeable. One way to demonstrate that they can be changed is to take a problem statement and examine it phrase-by-phrase for hidden assumptions. The following is an example:

How to *improve* the performance of our *current engineering staff in managing projects.*

Table 27.4. An Excerise in Nonlogical Stimuli.

For this exercise, you will need paper and pencil and a dictionary.

1. Write down as many uses as you can think of for a piece of chalk.

2. When you can think of no more ideas, let your eyes wander to some object in your range of vision, which has no immediate connection to a piece of chalk.

3. Try to develop new ideas stimulated by the object.

4. Now repeat stages (2) and (3) with a second randomly selected object.

5. Open the dictionary and jot down the first three nouns or verbs that you see.

6. Try to develop new ideas stimulated by these words in turn.

7. Examine your ideas produced with and without stimuli for differences in variety (flexibility) and total numbers (fluency).

The italicized words can all be examined. Should we try to improve the performance of our staff, or should we perhaps appoint project managers who are separate from the engineering staff? Is it our staff who are not performing through some innate problem, or is the system the cause of difficulty? Should the engineering staff be managing projects at all? Is it the management of projects that is the problem or are we doing the wrong projects in the first place?

Reversals

Sometimes the best way to do something is to not do it. By turning a problem upside down and examining the paradox that is created, one can sometimes see new approaches. For example, if a product has a weakness, try to make it a strength, as in the case of NyQuil™, which was a great cold remedy, but had one drawback—it made the patient sleepy. The question—how to turn that disadvantage into an advantage. The answer was to sell NyQuil as a nighttime remedy that would actually help the cold-sufferer get some sleep.

A food low in nutritive value becomes a diet food. A glue that wouldn't stick permanently was the key to making Post-it™ note paper. (The idea was rejected initially. Who needs such a thing? It was a number of years before 3-M decided to market the product, and it is hard to imagine the world without Post-it paper now. In fact, in conjunction with a white marker board, Post-it paper is a great tool for project planning.)

Linear Techniques to Generate Ideas

For almost every problem, we might begin by asking, "How can we do that . . . ?" How can we, for example

☞ Develop a new product—or an idea for a new product?

☞ Build the new bridge most effectively and efficiently, so we make best use of our resources and make the most money?

☞ Put together a new training program for our lab technicians?

Linear techniques for generating ideas can be extremely useful in this area. An excellent reference is the book by William C. Miller (1986), *The Creative Edge,* which is cited in the reading list. Miller lists ten so-called linear techniques for organizing known information to help you see your problem from different angles.[1]

Matrix Analysis

> Nothing is more dangerous than an idea when it is
> the only one you have.
>
> —Emile Chartier

Matrix analysis is ideal for developing new product ideas. Suppose you wanted to investigate all possibilities for marketing training programs. You might then have a grid (matrix) that looks as follows:

Delivery Method	Client Groups			
	Managers	Engineers	Trainers	Retirees
Seminars				
Cassettes				
Videos				
Films				
Home Study				
Workshops				
Computer				

Each box in the matrix (each intersection as it is called) represents a place to look for new innovations.

Morphological Analysis

If you want to consider more than one or two variables, the matrix is not a very effective approach. Morphological analysis is probably better. As Miller says, this is a fancy title for a simple way to generate solutions to problems that have many variables to consider. For example, to continue with our training programs, we might have to consider

☞ delivery method

☞ course content

☞ audience

☞ location

Some of the topics that might fit into these categories are shown below.

Delivery Method	Content	Audience	Location
video	technical	college student	local
audio	behavioral	factory workers	foreign
workbooks	reading	managers	different state
films	writing	farmers	traveling
seminars	coping	housewives	samestate
satellite	agriculture	schoolchildren	shipboard
computer	computer science	professionals	nationwide
mail	medical	paramedics	

Once the table is prepared, a single variable in each column is circled and the possibilities are considered. For example, suppose we circled *seminars, coping, factory workers,* and *nationwide.* The immediate ideas that come to mind are seminars designed to help workers cope with being laid off during the recession. They might need help with the feelings of frustration and self-doubt that invariably accompany such situations, as well as training in how to prepare a resume, conduct themselves in an interview, and conduct a job search.

Attribute Listing

If you want to improve a procedure, product, or process, you might write down all the attributes or components and see how you can improve any one or all of them.

For example, suppose we want to improve the project management process itself. It has the following attributes:

☞ schedule

☞ overall plan

☞ project team

☞ form of organization

☞ control system

☞ project manager

If we examine each of these attributes, we might ask how it can be improved. For example, how do we improve our scheduling methodology? Is our form of organization optimum? Is the control system functioning to keep the project on track?

Alternative Scenarios

The two primary ways of exploring possibilities for the future are hypothetical situations and alternative scenarios. With hypothetical situations, you make up something and develop a solution for it. "If a certain set of conditions existed, what would I do?" To which of these conditions are we most vulnerable? What can we do about those vulnerabilities?

Alternate scenarios are more comprehensive than hypothetical situations. They are qualitatively different descriptions of plausible futures.

When long-range planning is based on a single forecast of trends, there is a big risk of "betting the farm" on that single forecast. Thinking through several scenarios is less risky, and frees one to take more innovative actions.

Scenarios are developed specifically for a particular problem. First, a statement is made of the specific decision that must be made. Then one identifies the major environmental forces that might impact the decision.

These forces might include technology, social values, economic growth, tariffs, and so on. Now a scenario is built based on the principal forces. To do so, use information available to you and identify those plausible and qualitatively different possibilities for each force. Assemble the alternatives for each force into internally consistent "stories," with both a narrative and table of forces and scenarios.

Forced or Direct Association

This approach is similar to nonlogical stimuli, which was introduced in the previous chapter. New ideas can be generated by putting together two concepts that seemingly have nothing in common. For example, if you were trying to understand how to improve the performance of a work group, you might ask, "How is this group like a roller coaster?" The following list might result:

☞ tracks

We stay on the tracks, but they just go up and down and around in a circle. All we seem to be doing is making ourselves sick.

☞ cars

The cars are designed to keep you from falling out. Maybe we aren't taking enough risks.

☞ speed

We aren't going anywhere, but we're getting there pretty fast.

☞ control

The person controlling the roller coaster just started it going and went on a break. Who's in control here, anyway?

With these ideas, you might identify ways to respond to the situation.

Design Tree

Another word for design tree is "mind map." These have been used by a lot of people to illustrate associations of ideas. For example, one author has a book on writing that makes use of mind maps. You begin by writing a single word—representing the issue you want to deal with—then draw a circle around it. Next list all the ideas that come to you, connect them to the first word with lines, and continue by examining each new word in turn for the ideas it might trigger. I used the word *transportation* to illustrate the approach. See Figure 27.1.

Endnote

[1]See Miller (1968) page 66, for his listing.

Figure 27.1. Design Tree for Transportation

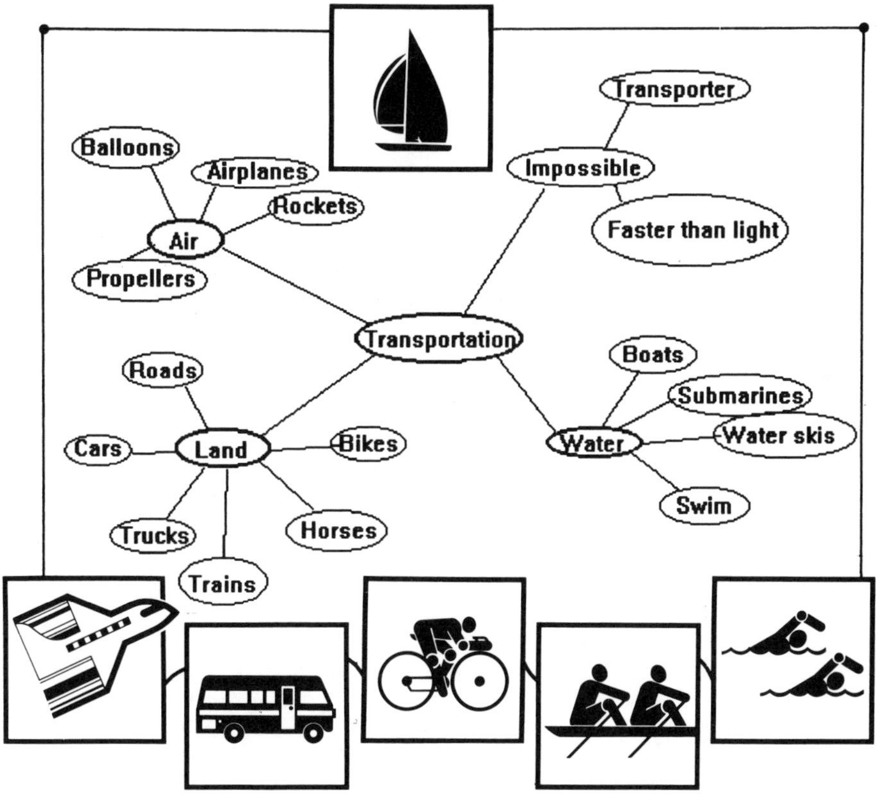

Chapter 28

Managing Decision Processes in Project Teams

How to Make Decisions in Teams

When the subject of making decisions comes up in group discussions, there are usually people from two camps. In one camp are those who say, "*Somebody* has to make a decision and the group has to live with it." In the other camp are those who say, "They should do it by *consensus.*"

In spite of all the concern expressed during the past decade about the negative impact of authoritarian management, the first camp feels that *someone* must be in charge, or chaos results. The other camp, influenced by the trend toward participation or democratization, believes that *everything* should be decided by consensus.

To be blunt, neither side is entirely right.

For all decisions to be made by the group leader is authoritarianism at its best, while making all decisions by consensus is a gross waste of time for a group.

Naturally, then, the question is, just what is the correct procedure?

Problems and Pitfalls

Jerry Harvey (1988) uses a story to illustrate the pitfalls of group decision-making. A family is sitting around on the porch of their home in Texas one Sunday, and they are bored. Someone asks what the others would like to do.

"How about if we drive into Abilene and have lunch at the cafeteria," a member of the group offers.

There is very little discussion, but after awhile, the entire group piles into the family's old Buick, which has no air-conditioning, and they drive about 90 miles to Abilene. The temperature is nearly a 100 degrees in the shade, so we know they must have been a bit weather-beaten by the time they arrived.

They have lunch in the cafeteria. The food is mediocre. Then they walk around the streets of Abilene for a while, but there is nothing to do, so they are bored in Abilene.

Finally, they pile back into the old Buick and sweat another 90 miles back home.

They get out of the car, start dragging themselves back to the house, and Ma says, "Boy was that ever a waste of time."

"I thought you really wanted to go," Pa says.

"No, I didn't want to go," Ma says, "I just went because the rest of you wanted to. I'd just as soon have stayed here and fixed lunch myself."

With that, they took a poll and found that *nobody* really wanted to go to Abilene—not even the person who suggested it. It was only an idle thought.

What happened?

They were trapped by the prevalent assumption *silence means consent*. As Harvey says, this is the *false consensus* effect. It is a flaw in how we manage dissent, not how we manage agreement. We often do not take steps to bring out dissent, and conclude that there is none, since no one spoke up. Then we find that the group really was not for the idea, and often they won't support it.

A related problem was labeled *groupthink* by Dr. Irving Janis (Janis & Mann, 1977). He cites some well-known examples of this. One was the Bay of Pigs invasion of Cuba, which President Kennedy had to deal with early in his presidency.

When he learned that Cuban exiles were preparing to invade Cuba, he asked his advisors for opinions on the operation. They agreed that the invasion should proceed.

But, there was one dissenter—Arthur M. Schlesinger, Jr. During a break in the conversation, Bobby Kennedy, the President's brother, called Schlesinger aside and asked why he was "rocking the boat." After all, it was clear, he said, that President Kennedy wanted to go ahead with the invasion.

Note the expression "rocking the boat." People often consider a dissenter in a group to be a boat-rocker or a non-team player. However, experience shows that, if teams are to be really effective, members must feel free to express their true feelings about issues. Nevertheless, there are pressures exerted sometimes for all members to conform to the majority position.

To continue, as a member of a new team, your first order of business is survival. Schlesinger realized that continued disagreement on his part would damage his standing in the group, so he shut up.

The invasion of Cuba was a disaster. Later, president Kennedy realized what had happened and told his cabinet members, "I don't want a 'yes-man' cabinet. When I ask for advice, I want to know what you really think."

The false consensus effect can operate on its own, or it can be part of the groupthink phenomenon. When groupthink is operating, the leader expresses a preferred course of action and nobody challenges it, so he or she assumes they all agree. Then they wind up in Abilene or the Bay of Pigs when they don't want to be there.

Decision-Making as Group Process

As these stories illustrate, decision-making is one of the processes in the functioning of a team that must be handled correctly or the task performance of the entire team may suffer. Fortunately, the question of how to handle group decision-making can be answered with some precision, thanks to the work of several experts on group problem-solving and decision-making.

In a book dealing with group decision-making and problem-solving, industrial psychologist Dr. N. R. F. Maier (1963) showed that decisions have two possible components. One of those has to do with the merits of the choice being made. Maier called this component a *quality* component, which means that there is some measurable (or quantifiable) way to say that one choice is better than another. Unfortunately, the word *quality* has so many connotations today that to use it in this sense only leads to confusion, so I have chosen the word *merit*. If one choice has a measurable advantage over others, it has greater merit.

The second attribute which the decision may have is *acceptance*. Members of the group may have feelings about the choice, which will affect their acceptance of it, and if they do not accept it, successful implementation is unlikely. The reason is that people who do not accept a choice are generally unwilling to support it.

This was confirmed for me by the superintendent of a county school system, who told me that when her group of principals vote on a course of action, she later finds some principals not supporting it. When she comments on their lack of support, they say, "Well, you may remember, *I didn't vote for it either!*" This shows why voting is a bad way to make group decisions if the support of all members is needed to make the course of action work.

The third factor that affects how decisions are made is *time*. When there is a tight deadline to be met, some sort of time-reducing strategy must be employed.

It is possible for a decision to have **only** a merit component or **only** an acceptance component, but such absolutely "pure" types are rare. Most decisions have some degree of both components present. The most frequent mistake that I have seen made by group leaders is to forget the acceptance attribute. I believe that they often think the decision is only a merit issue, since most decisions have some merit component.

Table 28.1 presents the rules for handling group decisions, as a function of how "strong" each of the attributes happens to be. When merit is the most important component, the decision is made by the person best qualified to do so. That does not necessarily mean the *leader*. For example, a technical or engineering decision should be made by the expert on that topic.

When acceptance is dominant, group consensus is generally the approach that should be used. When both merit and acceptance are involved in a decision, the style to be used is called *consultation*. An expert deals with the merit component, presents the group with equally acceptable options, and allows them to have input into the final choice.

I believe it is also appropriate to use groups to decide issues in which no single person has all the facts, expertise, or whatever. The Bay of Pigs situation is a good example of the need for the entire group to be involved.

Table 28.1. A Guide to Decision-Making Style (Based on Maier)

DECISION TYPE—suggests—DECISION STYLE

$$\frac{M}{A}$$

Autonomous

Merit of the decision
is more important than
its acceptance

The decision is made by
individual most competent
to do so.

$$\frac{A}{M}$$

Consensus

Acceptance of the decision
is more important than its
merit

The decision should be made
by those who must accept it.

M-A

Consultation

The merit and acceptance
of the decision are almost
equally important

The decision is made by the
most competent person, while
keeping those affected involved
and informed.

Vroom and Jago's Procedure

The rules developed by Maier have been researched extensively by Victor Vroom and his associates. They have developed an elegant model to guide leaders through the process, using decision trees to specify the appropriate decision method. The interested reader should consult the book by Vroom and Jago (1988).

Managing Group Consensus

Group consensus is difficult to achieve. The precise meaning of the word *consensus* suggests that "everyone agrees," and getting everyone in a group to agree to *anything* has such a low probability as to be considered impossible. It may be for that reason alone that group leaders sometimes decide that they should either make decisions themselves or that they should fall back on majority rule.

For practical reasons, *consensus* must have a different meaning than "everyone agrees." Naturally, in arriving at a decision, polling of group members is generally used to determine majority position, but the next step is critical. In order to be able to say that a consensus exists in the group, the minority members must *all* be able to say individually:

"While I do not entirely agree with the majority of you, I understand your position and I feel that you have given me a fair hearing. I am fully willing to *support* the majority preference."

The key to success is that each member be fully willing to support the majority position. If each member can say this, then you have as close to consensus as you are likely to obtain with a group.

However, if there is a member who is unable to support the majority position—especially if there is a member who is saying, "Not only can I not agree with the majority, but I am not even willing to support that choice," then you need to look at other options. Those options are as follows:

1. Find another course of action with which the dissenter can agree, as well as the majority of members. There is always more than one way to do anything.

2. Continue the discussion. Try to understand the dissenter's position. Perhaps he or she is right and the majority is wrong.

3. Another option is to have the person "sit this one out" or, if that is not possible, you may occasionally have to remove the person from the team if his or her commitment and support are critical to success. The danger with this option is that people may conclude that dissent means you get kicked off the team. It should be used sparingly. I am convinced that almost always you can find a way to achieve option one.

4. The group continues to argue its case with the dissenter until he or she is convinced to go along with the majority. The risk of doing this is that group pressures can gain outward compliance without inner commitment, and later, when support from the person is critical, it will not be forthcoming.

How to Avoid Groupthink

As was pointed out at the beginning of this chapter, groups sometimes accept a position taken by their leader, in spite of evidence that the leader may be wrong. Whatever the cause of such a phenomenon, groupthink may lead to dire consequences, and leaders are advised to try to prevent it if possible.

Janis suggested that groupthink can be avoided if a specific procedure is followed in reaching group decisions. The following steps summarize Janis' procedure.[1]

1. The leader should avoid expressing a position on the issue until after all group members have aired their own views.

2. Alternative courses of action should initially be offered with no evaluation allowed.

3. Once alternatives have been proposed, every member of the team is expected to play the role of "critical evaluator," looking for flaws in a course of action.

4. A choice should be made through consensus if possible, avoiding such approaches as majority voting. The definition of consensus which was presented above should be used.

5. The group should "sleep on their decision" overnight, and reconvene the next day for a last chance to discredit the choice. If a member has had new thoughts he or she is encouraged to express those concerns.

Naturally such a procedure requires considerable time, and should not be employed unless the issue is of significant importance to the entire group, or unless the entire group is absolutely necessary to deal with an issue. Otherwise, the procedures offered by Maier should work well for most issues.

Endnote

[1] Adapted from Janis and Mann (1977).

Section Ten

Improving Communications in Projects

Chapter 29

Improving Communication in Project Environments

Improving Interpersonal Communication

A significant aspect of a project manager's job is to communicate to his or her people the objectives or purpose of the project. When I am consulting to organizations having problems with project management, I ask managers, "Are your people all clear on the goals and objectives of the project?" Usually I am assured that they are.

Then I ask members of the project team if they are clear on the goals of the team, and I get a contradictory answer. Or I sometimes just become aware that they are not clear through discussions with them.

In many cases this happens because managers often communicate in one direction only. They do not take time to be sure that the person with whom they are communicating actually understood the message. Rather, they assume that everything was clear.

In fact, there seems to be a belief among managers that if they are fairly good at expressing themselves, then this is all that is needed to make them good communicators. To show the fallacy of this, once when I was in Mexico, my wife and I got into a cab to go to the airport. I spoke to the driver in Spanish, and he apparently assumed that I spoke and understood Spanish very well, because he launched into a long story about how bandits had tried to rob him and he had pulled a pistol from the glove box and scared them away, and the story was told with much feeling, laughter, and waving of hands in the air.

When we arrived at the airport, my wife wanted to know if I understood him, as she knew that my ability in Spanish was very limited. Of course, I did manage to get the drift of the story, but I am sure I missed many of the details. The point is, the driver assumed that I understood, simply because I was able to say a few things in the language.

So, too, we make this mistake with our followers. We speak to them. They nod their heads, grunt affirmatively, smile, and ask a few knowing questions. Clearly, they must follow us perfectly, yes?

Not so.

The terrible truth of this has been demonstrated for me in an exercise that I sometimes do in my seminars. I divide people into groups, then tell them, "For the next 15 minutes, you can do whatever you want to do, so long as you do it as a group and report back to the rest of us what you did."

They look at me with some confusion evident, and I simply repeat the instructions verbatim.

They then stand around for a moment trying to decide what to do.

"Anybody got any ideas?" someone asks.

Initial silence.

Then a feeble, "How about if we go out into the parking lot and count how many license plates are from each state."

That goes over like a lead balloon.

"Any other ideas?" asks the emerging leader.

"Whatever we're going to do, we better get with it," says someone who is looking at his or her watch nervously. "We only have 13 minutes left."

No one has any other ideas, so they divide the parking lot into zones, assign an individual to each zone, and then they count tags.

Other groups do similar things.

After they have given their reports, I ask how they felt about the clarity of the goal. They admit to some anxiety, though it is not too strong, since nothing great is at stake in a classroom setting. (The same would not be true back at work, where there are real penalties for failure.)

"Why didn't any of you come and ask me for clarification?" I ask.

"Well, we didn't want to appear stupid," someone says.

"Do you ever get fuzzy objectives back at work?" I ask.

"Is there any other kind?" someone quips.

"Well, when you get fuzzy objectives at work, what do you do?"

"We try to figure out what they really want us to do, and then we do it."

Incredible!

I have been dismayed by this finding for a long time, but I recognize it as part of Western culture. When students don't understand in class, their classmates or—worse yet—their teachers put them down. They are told they have asked a stupid question.

This is nonsense. The only stupid question is the one that isn't asked.

Further, teachers have not learned lessons from quality control, namely, that those students are their customers, and that the teacher's job is to meet the needs of those customers, and failure to do so is one of the great illnesses of our educational system. But that is another matter.

Suffice it to say that we are socialized to be poor communicators. Managers take for granted that they are understood and employees are afraid to ask for clarification because they don't want to appear stupid. It might cost them at appraisal time.

As a result, things are done wrong. Worse yet, *the wrong things are done!* Sometimes badly!

And, before you think I see myself as the grand communicator, who never makes errors, let me assure you that this is not the case. To illustrate, I once had an employee to whom I would give a job assignment and ask, "Do you understand what I want you to do?" He would answer yes and then go off and do something else. When I asked why he did something different, he would say, "You told me to." Then we would argue about what I (believed I) actually told him.

Finally one day I sat down with him and gave him an assignment. When I finished, I said, "Let me be sure I communicated to you what I want done. Tell me what you understand the job to be." When he did, he was completely off-target. Now I understood that when he said he understood, it simply meant he heard me.

I am not saying that he was at fault. It is possible that my mode of communicating was the cause of the difficulty. Perhaps my phrasing, the words I used, or some other element made it difficult for him to understand what I intended to convey.

It is easy to label the employee *dense* or a *poor listener*, but that does not solve the problem. I have learned two axioms that guide my thinking about—and my approach to—communication.

> **The meaning of a communication is the response it gets.**

> **Responsibility for communication rests with the *communicator*—not the other person!**

The first axiom says that the net result of a communication is the target person's reaction to it. It makes no difference what you *intended* to communicate, insofar as the result is concerned. What the person heard and understood will affect his or her response.

The second axiom says that the communicator must take responsibility for ensuring that the message *received* by the target person is the message actually *intended!*

I learned the importance of the second axiom from this experience with my subordinate. You must close the feedback loop between yourself and the other person to be sure the "message" got through (see Figure 29.1). Further, closing the loop cannot be accomplished by letting the person say "uh huh," or even, "I understand." You must solicit specific feedback, asking the person to use his own words, in order to be sure the message was understood.

Figure 29.1. Close the Loop to Communicate.

Naturally, some people will be offended by this. They may say, "You must think I'm dense." To those people, I simply tell my story and explain that the communication problem can be my own poor way of expressing myself, and that I cannot afford to have problems caused by a miscommunication, so this is my way of ensuring that I conveyed the message as intended.

This leads to the following rules for effective communication.

Basic Requirements for Communication

1. Know what *outcome* you want.

2. Decide to *whom* you need to communicate. Is it the entire group? One person?

3. What is the best mode in which to communicate? Written? Verbal? Both?

4. Have the sensory awareness to notice when you get the response that you want. Best of all, close the loop.

5. Acquire *flexibility*. Be able to vary your communication until you manage to get through to the other person. If you continue to repeat yourself using the same words, you are deadlocked.

The flexibility issue is important. Have you ever asked a teacher a question and he or she explained the subject to you again—using almost exactly the same words as were used the first time? If those words had worked the first time, you wouldn't have asked a question, but the person doesn't seem to know how to say it differently, so he or she repeats the script, like a tape recorder.

When someone does not follow you the first time, it is important to try to find a different way of saying what you just said.

Requirements for Effective Communication

Communication depends on having:

☞ A common culture. Cultural differences can come from nationality, social standing, part of the same country in which a person grew up, race, sex, and profession. There are cultural differences between people from the northeastern United States and the South. The same is true between California and other parts of the U.S. These differences can lead to miscommunication and conflict. And, since the United States is becoming increasingly multicultural, managers need to understand some of the differences so that they can deal with them.

Even professional differences lead to difficulty. We often use jargon and don't explain its meaning, thus leaving the other party in the dark. And, because of our reluctance to ask "stupid" questions, the other person may not ask what you mean by *PMOs*, which **clearly** means Preventive Maintenance Orders (to the person using the term).

A fellow recently told me a marvelous story about cultural difficulties resulting because of nationality. He was sent to Japan to participate in the opening of a new factory there. He was instructed in the importance of presenting his business card properly to his Japanese contacts, as this is a very significant ritual there. So, during the ceremony, he faced his Japanese counterpart, walked forward, bowed to the proper level, and held his business card out in front, as he had been instructed to do.

To his surprise, the Japanese gentleman, who was also bowing to him, whispered, "You're dead."

"I beg your pardon," the American whispered.

"You're dead," the man said again.

It was then that the American realized he was holding his card upside down, which in Buddhist culture means the person whose card is so displayed is dead.

Another story that he told me illustrates that cultural differences can lead to misunderstandings and "strange" behavior on the part of members of a project team. In most companies, job openings are posted to give all employees a fair and equal opportunity to move up. His company found that when they posted a job none of their Hispanic men were applying. They didn't understand why, since many of the postings would have been good opportunities for their employees.

When they investigated, they learned that to the Hispanic male, asking for a chance at a better-level job was considered demeaning.

☞ Common expectations. When expectations differ, communication suffers.

☞ Motivation to communicate. You will have a hard time "getting through" to someone who has decided he or she doesn't want to hear what you have to say.

Modes of Communication

Verbal: Verbal communication consists of the words we use. Experiments have shown that only about 10 percent of the *meaning* of a message is carried by the words in typical face-to-face conversation. Fully 90 percent of the meaning is carried by the nonverbal part. The verbal element is sometimes called the *digital* channel.

Nonverbal: Everything that is *not* words falls into this category. Included are facial expression; voice characteristics, such as tempo, tonality, volume; proximity of speaker to target person; gestures; body posture; and so on. Nonverbal communication is sometimes called the *analog* channel.

Because nonverbal communication must be *interpreted,* it is very subject to misunderstanding. The only way to be certain what someone's nonverbal communication means is to *check it out!* This is called *calibration* in neurolinguistic programming, a field of behavioral science developed by John Grinder and Richard Bandler in the late 1970s.

To see how meaning changes with changes in the nonverbal element, consider the following question. The italic word indicates where the speaker applies stress.

☞ *Can* you solve the problem?

☞ Can *you* solve the problem?

☞ Can you *solve* the problem?

☞ Can you solve *the* problem?

Note how the meaning changes with each change in the word stressed, even though the words themselves remain the same.

Symbolic: How we look to the other person. This includes our clothes, jewelry, car, hairstyle, and so on. Symbols tell others who we are and give them a cue as to how to communicate with us. If you are a manager, but don't *look like* one to the other members of your organization, then you won't be fully accepted in that role. Note that whether you look like a manager is a function of the norms that exist in your organization. If all managers wear pin-striped suits, dress shirts and ties, and you wear sport jeans, sneak-ers, and no tie, you will be out of sync with the culture of your organization.

Tactile: In U.S. culture, touching tends to be reserved to close friends and intimate partners. It is very risky to touch someone un-less you know them very well. Because this is not true in other cul-tures, it can lead to misunderstandings when people from different cultures interact.

Content and Relationship

Another aspect of communication that often causes friction is the *relationship* aspect of the message. Every time you communicate with someone, you convey a message—the content—as well as a definition of the way you view your relationship with him. If you offer a definition of *relationship* which is unacceptable to the person, she may reject the content. You sometimes hear people say, "It wasn't what the other person said, it was the way they said it that offended me."

For example, you ask an engineer a technical question about his work, and he thinks you are criticizing him. All you intended was to gain a better understanding of what he had done.

Or you ask how an individual's work is proceeding, and he is offended. "You're always looking over my shoulder, checking up on me!" he says angrily. "I guess you just don't trust me to do my job." And all you wanted to do was to show an interest in his work.

Relationships can be close-and-personal, distant, husband-wife, supervisor-subordinate, peers, and so on. Some of them we con-sider to be generally equal-status. Others we consider to be un-equal-status. Your peers are on an equal status with you, while your subordinates are not (speaking only about the status of an in-dividual in the traditional organizational hierarchy). However, your subordinates may resent it if you "come on too strong." We do not like it if people emphasize their higher status to us too strongly.

For this reason, we find that some people will resent being told directly to do something. For example, if a manager says to an engi-

neer, "Tom, run a stress analysis on this, and let me know the results," Tom may be offended. He may see his manager as being too
pushy. (Note that he may not, also—there is considerable variability
in how people view this issue.)

So his manager may find the directive more palatable to Tom if
he says instead, "Tom, how about running a stress analysis on this
and let me know the results." The message is the same. It is still a
directive. But Tom may not be offended by it because it is framed
as a *request*. We say it sounds more *polite*. Actually, it is more equal-
status, and is therefore less likely to rankle.

Language of Logic and Emotion

Engineers, programmers, and other technical people tend to be left-
brained, analytical, and logical. They like to deal with facts and
things. Nevertheless, that does not mean that they have no emotions. In fact, arguments over the *right* way to do something of a
technical nature often get very heated!

Left-brained people sometimes have difficulty dealing with
that emotional aspect of people. They would like them to be logical
and fact-oriented all of the time. But this is simply not how people
are, and that leads to another premise about dealing with people.

> **You must deal with people as they are, not as you
> would like them to be!**

It has always seemed to me to be unrealistic, and also inconsis-
tent, for a manager to tell a person, "Leave your emotions outside
when you come to work." Then in the next breath he wants the
person to be *excited or enthusiastic* about something, or he wants to
motivate him or her to do a job. Note the common root between the
word "motivate" and "emotion."

What such a manager is saying is, "Leave the negative emotions outside, but bring in the positive ones." Obviously, this is impossible. In fact, it is that emotional component of human beings

that separates them from machines and gives them their real strength.

Actually, the manager simply is trying to "cop out," as we say. He is trying to avoid dealing with emotion because he lacks the skills to do so. For that reason, engineers who become managers must sharpen their skills in dealing with emotions in people, since that comes with the "turf."

This leads to another premise about communicating with people:

> **The language of logic and the language of emotion are different!**

Another way to say this is that you cannot deal with facts when someone is upset. First you must get him calmed down. Then you can return to logic.

The single most important point to remember about dealing with someone who is upset is that he generally feels that you don't understand him. And you cannot convey to him that you do understand by simply saying so.

The best approach to follow is to simply reflect back to the person what you heard him say, along with your sense of what he is feeling. Here is an example:

Engineer: I really feel that you don't appreciate how hard I've worked on this project. I've been out here every evening until nearly midnight, busting my buns, and you seem to take it for granted.

Manager: Those are long hours, Tom. You say you feel that I don't appreciate the effort. Can you tell me what I did to make you feel that way?

Engineer: Yeah. You've never said you appreciate the long hours. All you do is come in here and ask when I'll be finished.

Manager: I see. Since I've never told you that I appreciate your hard work you have the impression that all I care about is pushing you to get through.

Engineer: Right.

Manager: I can certainly see why you would feel that way. I haven't been very sensitive to the hard work you've been doing. I'm sorry about that. I do appreciate it, and I should have said so sooner.

From here, the conversation can go in a lot of different directions. In any case, by this point, the engineer should be feeling less angry, and the manager can now move to resolve any problems using a logical approach. But until the engineer feels that his boss understands his feelings, logic will not work very well.

This approach is called *active* or *reflective listening,* as opposed to passive listening (characterized by a lot of head nodding, uh huhs, etc.). Note that the manager reflected back what she heard the engineer say—*in her own words.* She did not simply parrot back the exact words of the other person.

Further, she tried to empathize with the person by listening for his feelings, and she commented on those.

Note also that the manager admitted his or her error and apologized for it. Only an arrogant manager refuses to admit mistakes, and in doing so, alienates the very people on whom he or she depends to achieve success.

Active listening skills are a necessary part of every manager's "tool kit" for dealing with people. It requires practice, however, since we have learned to simply respond to the other person's comments with a rebuttal. Note also that the approach is valuable just as a general approach to ensuring that you heard what someone else has said.

Section Eleven

Special Topics in Project Management

Chapter 30

Training and Educating Project Managers

The Need for Training

There seems to be a prevailing belief in the United States that if you are good at *doing* something, then you can manage other people doing that same work. Since 1981, when I entered the training and consulting business, I have talked to thousands of people who were put into management positions and given no training in how to manage. In fact, the same thing happened to me, and I know first-hand the fallacy of the implicit assumption.

To make matters worse, we don't even require people to gain experience in their field before we start making managers of them. I have asked several hundred engineers, who graduated within the past couple of years, "How soon after you started working as an engineer did you get your own project to manage?" The answers range from "within two hours" to time frames of a couple of months.

Yet very few engineering schools are offering training in project management, and even if they did, the courses would probably not be very relevant to an undergraduate, so that by the time they go to work, most of the material would be forgotten. In fact, the few engineers who have received any training at all have only had a course in scheduling.

I personally believe that this lack of training is one of the major causes of project failures. So this chapter is intended to tell readers what kind of training they need, and where to get it, beginning with noncredit training and some do-it-yourself methods, and finally followed by graduate programs in project management.

Skills Needed by Project Managers

Following is a list of the primary areas of knowledge and/or skills needed by project managers.

◆Planning	◆Decision-making	◆Problem-solving
◆Conflict management	◆Setting goals	◆Analyzing data
◆Negotiation skills	◆Leadership skills	◆Oral communication
◆Written communication	◆Interviewing	◆Coaching/counseling
◆Group dynamics	◆Team building	◆Quality function deployment
◆Total quality management	◆Scheduling methods	◆Concurrent engineering
◆Earned-value analysis	◆Listening skills	◆Time management

Noncredit Training in Project Management

There are many universities, individuals, and other organizations that offer seminars and workshops in project management. These range in duration from one day to a week. The one-day programs naturally provide only an overview of project management, and most participants find that they have only achieved an understanding of concepts upon completion of the course. It takes at least two days to cover the fundamentals of planning, scheduling, and control in enough depth that students can apply what they have learned. Programs are best when they contain some group and individual exercises, which permit students to test their understanding of principles through application to a classroom problem. In addition, case studies are helpful for allowing group discussion and testing one's application of principles in hypothetical situations.

If you are trying to locate a seminar and you live near a university, you might contact their short-course or continuing education department to find out if they offer project management programs. Following are some providers of such programs:

Institutions/Consultants:

American Management Association	New York	518-891-0065
Applied Business Techology	New York	212-219-8945
Centre for Management Technology	Singapore	011-65-345-7322
International Institute for Learning	New York	212-909-0557
Lewis Consulting & Training	Vinton, VA	703-890-1560
QualityAlert Institute	New York	800-221-2114
Quentin W. Fleming	Tustin, CA	714-731-0304
Richard Villoria Associates	Santa Monica, CA	310-207-3862

Universities:

Arizona State University	Tempe, AZ	602-965-3441
Brunel University	London	895-27 40 00

Clemson University	Clemson, SC	803-656-2200
George Mason University	Herndon, VA	703-733-2800
Michigan State University	Lansing, MI	517-353-8711
University of Central Florida	Orlando, FL	407-275-2446
University of Tulsa	Tulsa, OK	918-592-6000
University of Wisconsin	Madison, WI	608-262-2155
Washington State University	Olympia, WA	206-593-8575

Self-Study Programs

Video courses on project management are available from PM Tools, P.O. Box 2777, Ann Arbor, MI 48106-9887. Phone 313-996-0529. They also carry a number of books on project management.

Graduate Programs in Project Management

At least three U.S. schools are offering Master's degrees in Project Management at the time of this writing (October 1992). I believe that more will do so in time, as project management becomes recognized as a very specialized discipline. For information on the programs contact the following schools. (Readers are invited to send me information on other programs for inclusion in future editions of this handbook.)

School: **Golden Gate University**
Location: San Francisco, CA
Program: M.S. Degree in Project and Systems Management (offered through the School of Technology Management)
Contact: Director of Admissions, Golden Gate University, 536 Mission Street, San Francisco, CA 94105. Phone 415-442-7000.

School:	**Keller Graduate School of Management**
Location:	Main campus is in Chicago. Satellite campuses are located in Lincolnshire, Illinois; Schaumburg, Illinois; Orland Park, Illinois; Downers Grove, Illinois; Kansas City, Missouri; Milwaukee, Wisconsin; Phoenix, Arizona; and Mesa, Arizona.
Program:	Master of Project Management
Contact:	**Chicago/Loop Center,** 10 S. Riverside Plaza, Chicago, IL 60606. Phone 312-454-0880.
	Chicago/North Suburban Center, 25 Tri-State, International Center, Lincolnshire, IL 60069. Phone 708-940-7768.
	Chicago/Northwest Suburban Center, 1051 Perimeter Drive, Schaumburg, IL 60173. Phone 708-330-0040.
	Chicago/South Suburban Center, 15255 S. 94th Avenue, Orland Park, IL 60462,. Phone 708-460-9580.
	Chicago/West Suburban Center, 1101 31st Street, Downers Grove, IL 60515. Phone 708-969-6624.
	Kansas City/Downtown Center, City Center Square, 1100 Main Street, Kansas City, MO 64105. Phone 816-221-1300.
	Kansas City/South Center, 11224 Holmes Road, Kansas City, MO 64131. Phone 816-941-0367.
	Milwaukee Center, 330 E. Kilbourn Avenue, Milwaukee, WI 53202. Phone 414-278-7677.
	Phoenix/Northwest Center, 2149 W. Dunlap Avenue, Phoenix, AZ 85021. Phone 602-870-0117.
	Phoenix/East Valley Center, 1201 S. Alma School Road, Mesa, AZ 85210. Phone 602-827-1511.
School:	**Western Carolina University**
Location:	Cullowhee, North Carolina
Program:	Master of Project Management
Contact:	James W. Pearce, Director of Graduate Programs in Business, Western Carolina University, Cullowhee, NC 28723. Phone 704-227-7401.

Course Requirements for Master's Degrees

Following are course summaries for the M.P.M. programs offered by Keller Graduate School and Western Carolina University. The curricula at other schools will no doubt vary, but this will give some idea of what a project manager is expected to know in order to receive a graduate degree.

Keller Graduate School:

Core courses (all four required)

AC 501	Financial Accounting Foundations
AC 503	Managerial Accounting
GM 533	Applied Managerial Statistics
MM 522	Marketing Management

Advanced required courses

PM 586	Project Management Systems I
PM 587	Project Management Systems II
PM 593	Project Leadership & Team Management
GM 588	Managing Quality
PM 557	Project Cost Management
PM 598	Contract and Procurement Management
PM 610	Capstone Project

Advanced elective courses (any two required)

PM 562	Financing Business Projects
PM 594	Project Economics
PM 584	Managing Software Development Projects
GM 583	Operations Management
MM 572	New Product Planning and Development

Prerequisite skills course (one or both may be required)

| GM 400 | Foundations of Managerial Mathematics |
| GM 410 | Business Communications |

Western Carolina University:

Core Courses, 15 hours, all required

ACCT 651	Managerial Accounting
ECON 607	Managerial Economics
FIN 601	Financial Management
MGT 605	Organizational Behavior and Analysis
MKT 601	Marketing Management

Advanced Competency Courses, 18 hours

MGT 670	Project Management Systems
MGT 672	Human Resource Management in Projects
MGT 674	Project Contract and Logistics Management
MGT 676	Specialized Project Management Techniques I
MGT 677	Specialized Project Management Techniques II
MGT 678	Project Management Organization and Policy

Foreign Language Training

For project managers who are involved in international work, learning a second language may be necessary. There are basically two approaches, classroom and self-learning. Classroom programs usually involve intense "doses" of training over perhaps a one-month period. While they are effective, they are also expensive, typical cost being in the thousands of dollars.

The other way to learn a language is through the use of tapes or records. A number of providers exist, and if you have the self-discipline, this method works. However, most programs rely on rote learning through "overlearning," as it is called, which means that you repeat, repeat, repeat, ad nauseam.

A much more effective approach was developed by Dr. Paul Pimsleur. He based his method on a principle from learning theory, which is that learning is more effective if a stimulus is presented to the learner at *unequal intervals,* rather than equal ones. Dr. Pimsleur developed his courses using this method, which he called *graduated interval recall.* As it turns out, the most effective approach is to present the stimulus in intervals of 2, 4, 8, 16, 32 time units. (The numbers given here are examples. For exact intervals used, consult Heinle and Heinle.) So, for example, if you were trying to learn how to say "potato" in German, you would be presented with the German word, which is *kartoffel*. About two minutes later, the instructor would ask, "Do you remember how to say 'potato' in German?" You would be given a chance to respond, and regardless of whether you got it right, the native speaker would say the word *kartoffel*, which would either verify your correct response or provide a memory jogger for you. Then about four minutes later, the same steps would be repeated. This would be done again in eight minutes, then sixteen, and again on a later tape, so that by the time you have had the stimulus word presented at least five times, in graduated intervals, it has gone from short-term memory to long-term memory. In other words, you *know* the word for potato now.

What this means is that the method makes learning virtually painless. You can work on the language in your car, on a plane, etc., because the instructions are given in your native language, making a book unnecessary. (In fact, a book is undesirable. A language is auditory, not visual, and it is far more effective to learn entirely through the ears than through the eyes. Once you can speak the language, reading and writing are easily mastered.)

Tapes using the Pimsleur method can be ordered from the producer, Heinle and Heinle Enterprises, 29 Lexington Road, Concord, MA. Phone 508-369-7525; 800-628-2597; FAX 508-371-2935. Tapes will ultimately be available in three levels, indicated by Roman numerals. They are available in the following languages (note that those in development are specified as such).

Arabic I	Chinese I	English I (for speakers of Spanish)	French I, II, III*
German I, II	Greek I	Hebrew I	Hindi I*
Italian I	Japanese I, II*	Polish I*	Portuguese I, II*
Russian I, II*	Spanish I, II, III*	Ukrainian I*	

* Forthcoming title (as of October 1992. Check with publisher for availability).

Putting It All to Work on the Job

Developing one's skills and abilities is a lifelong process. The process is most effective when it is self-directed and active. Learning by simply behaving like a sponge is not the most effective strategy for an adult. Rather, an active, seeking-out method has been found to be the best.

There is one barrier to learning behavioral skills that you should be aware of. That pitfall is your employees. If you start managing them differently, they are going to feel uncomfortable, because they can no longer predict you. To deal with their discomfort, they will try to pull you back into the old patterns.

In order to minimize the impact of your own subordinates on your new behavior, it helps to discuss with them the changes you want to make. This will keep them from experiencing such an abrupt change in your behavior.

Ask for their support and suggest that they discuss their feelings with you when they notice new behavior on your part. One of the best ways to neutralize resistance is to talk about it!

Developing New Skills

As a method of developing your skills in project management, the following procedure will help you deal with ongoing self-development.

   ☞ Identify the skill(s) you wish to improve or acquire:

   ☞ Assess your present level of ability using the scale below—draw a *circle* around the number that represents your **present** level:

The skill is . . .

Absent	Low	Moderate	High	Very High
0	1	2	3	4

☞ Decide what level you **want** the skill to be, and draw a *box* around the level on the scale above.

☞ List below what are some of the resources available to help you improve your skills in the area identified.

☞ Next, engage in the learning experience you identified above.

☞ Finally, re-assess your skills to determine if the learning objective was met.

Chapter 31

Sociotechnical Systems and Project Organization[*]

Traditional Project Organization

Over the years, a number of structures have been tried for organizing projects. Some of these are illustrated in Figure 31.1. At present, only two of the structures are widely used, those being pure hierarchy (sometimes called pure project form) and matrix. In fact, there

---

*   I would like to acknowledge the contribution of Michael C. Thomas, Ph.D., to my thinking in writing this chapter.

Figure 31.1. Some Project Organization Structures.

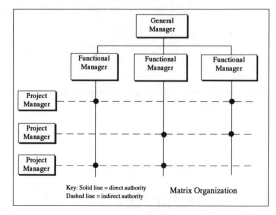

Key: Solid line = direct authority
Dashed line = indirect authority

Matrix Organization

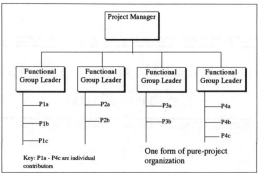

Key: P1a - P4c are individual contributors

One form of pure-project organization

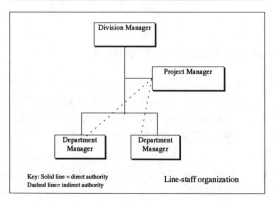

Key: Solid line = direct authority
Dashed line= indirect authority

Line-staff organization

seem to be a number of writers who almost consider matrix to be synonymous with project organization, though in practice this is by no means true. Many projects are still organized in hierarchical form when speed is of the essence in product development. Some of these are called *skunkworks* projects.

Both forms of organization have strengths and weaknesses, and these are described in another chapter of this handbook, so they will not be repeated here. Rather, this chapter examines project organization in the light of sociotechnical systems design principles, and offers some observations and recommendations for organizing large projects based on those principles.

The recommendations offered are not the cookbook type, however. That is, we do not have a step-by-step, how-to-do-it approach worked out. What we recommend is more in the nature of experiment, experiment, experiment!

That is not very comforting to those who want nice, tidy prescriptions, but it is the best we can do, given the state of the art today in organization design. My primary purpose in offering this chapter is to make readers think about the issues, so that at least an *awareness* of the complexity of the problem may be increased. We used to call this *consciousness-raising*. It is my belief that, by being aware of the issues involved, project managers will at least be able to avoid some of the problems of traditional forms of organization and perhaps invent some solutions.

What Is a Sociotechnical System?

The term *sociotechnical system* was coined by Eric Trist (Weisbord, 1987) to identify systems that are combinations of human and technical components. A system is characterized by four basic elements, as shown in Figure 31.2. It has *inputs, outputs,* a *process* that converts those inputs to outputs, and a *feedback* mechanism to regulate the transformation process.

We also have the term *open system,* which means that the system is open to interaction with its external environment, which of course suggests that the external environment will have effects on the system performance. A closed system, on the other hand, will not experience such effects.

Figure 31.2. A Feedback System.

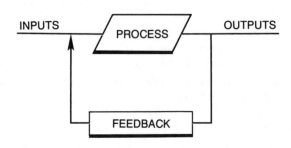

In a *sociotechnical system*, the process, inputs, output, and feedback elements are all combinations of people and technical "things" such as computers, manufacturing equipment, telephones, and so on. The basic gist of sociotechnical systems design is that there must be a joint optimization of the social and technical elements for the system to function at optimum levels. If only the technical components are optimized, then the social component may suffer.

John Naisbett has referred to our society as a "high-tech, low-touch" system, because technology has become so dominant that human interaction at the "touch" level has been reduced. We inter-

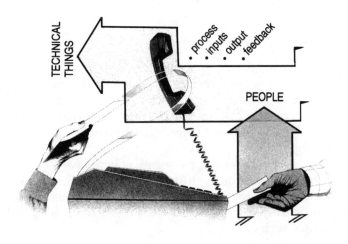

act by phone or computer, often not dealing with people face-to-face, which is a fact causing some grief to individuals who need that human interaction.

There is a third component that is actually a part of the social and technical systems, and which every organization contains, and that is its *reward* system. This component is a major influence on organization performance. As shown in Figure 31.3, the three interact, so that changes in one may affect the other two. For example, a change in technology will affect both the social and reward components. Introduction of computers changes the way people interact, which affects both the social system and the reward system, at least for those individuals who derive significant rewards from interacting with other people. In addition, some individuals will find working with the new computers to be rewarding, while others find it boring, threatening—in short, *un*-rewarding.

A change in the reward system may affect the social system as well as the way in which the technical system is employed. If people are rewarded for making better use of the new technology, then it should affect that utilization.

In short, there must be a joint optimization of all three components if an organization is to operate optimally, and that is the thrust of sociotechnical systems design.

Working Premises of this Chapter

This chapter is based on a number of premises that may be argued. However, I base them on the current body of evidence from studies and real-world organization events.

Need for Continuous Improvement

The quality movement of the 1980s seems to have convinced most people of the need to continuously improve organizations. As Dr. Edwards Deming, one of the most widely respected gurus of quality has said, there are two kinds of organizations—those that are improving and those that are dying. An organization that is standing still is dying, it is just that no one knows it yet. The reason is simple: the world is dynamic, and one's competitors are certainly improving their performance, so if you stand still for long, pretty soon your competition has left you behind.

Interestingly, many people do not have a good understanding of their competition and what they are doing, and so become com-

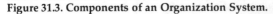

Figure 31.3. Components of an Organization System.

placent. I once asked some fellows from the sanitation department of a county government once if they had any competition.

"Nope," was the positive answer.

"You better wake up," someone in the audience told them. "If a private garbage collection agency bids a lower price to the county than you guys cost them, you'll be out on the street."

I agree completely with that position.

We also know of numerous organizations that have failed to see the impact of critical events on themselves until it was too late. Some have only one customer—in most cases, the military, and when that customer quits buying, they are in trouble. Others fail to realize how a major development in new technology will affect them, again, until it is too late.

The American auto industry has had to make major changes as a result of Japanese competition. At one time it required as much as a week or more to change over a production line to begin manufacturing a new model car. The Japanese got it down to hours, and American industry was forced to do the same.

Likewise, it used to take six to eight years to develop a completely new automobile design. The Japanese got it down to three years. You can't compete with someone who brings out a new model in three years when it takes you six years.

In fact, speed is almost the name of the "game" today. Tom Peters has emphasized this in a film that he titled *Speed Is Life*, in which he tells about some companies that managed to shorten their development times considerably through the use of cross-functional teams, and other approaches.

Another factor that enters into the equation is the coming labor shortage. Demographic forecasts for the United States are that by the year 2000 our population will grow by 12 million, while 14 million new jobs will be created. In addition, we will have a shortfall of some 565,000 engineers. Because of the 1991-92 recession, those figures may be off a bit, but one thing is certain, growth of jobs will eventually outpace the growth of population, so that every organization must find ways to *do more with less*. Thus the need for improvement of processes.

During the past decade, much of the effort to improve organizations has been aimed at manufacturing. Only in the past few years have people begun to realize the importance of improving organizations *across the board*. In fact, we seem to have reached a point in which reducing labor costs through improvements in human productivity may not be all that useful. Peter Drucker has argued that we now need to focus on the application of capital, as the largest portion of product costs are now in that area.

However, we still have a lot to gain through the improvement of performance in administration, support, engineering, and project teams. The costs to develop new products have reached staggering dimensions, with labor costs being in the range of $40 to $100 per hour. Because of those high costs, companies must either charge high prices for products or sell large quantities of them just to recover their investment, which has made the development of some prodcts too expensive, and those markets have gone to companies that have lower hourly costs for development.

The Impact of Management on Projects

Whatever happens or doesn't happen in a project occurs because management either wants it to happen or *permits* it to happen. The first part of the premise will be accepted by most people. The second, that management permits undesirable things to happen in a project may raise some defensiveness. Nevertheless, with the exception of acts of nature, it is management's responsibility to be monitoring project work closely enough to anticipate and address effects that might have a negative impact on a project, and failure to do so is to permit the impact through neglect.

Project Organization Is a Sociotechnical System

Based on the description of sociotechnical systems given at the beginning of this chapter, it is clear that all project organizations are sociotechnical systems. They employ technology and people, and have a reward component, all of which must be jointly optimized for the organization to function optimally. Sociotechnical systems design methods, therefore, should be applied to projects.

To date, the most visible applications of sociotechnical systems design have been in self-directed work teams and other job designs. As was mentioned earlier, this has restricted the application largely to manufacturing, but it is believed that the application can just as readily be made to project organization.

This chapter, then, examines what sociotechnical systems design principles have to offer in the organization of projects—especially product development projects, though it is believed that the principles can be applied to any kind of project team.

Sociotechnical Systems Design of Project Organizations

Involvement of All Members of the Organization

One of the tenets of the quality movement is that the person closest to a job is likely to be the most competent person to improve it. Whether this principle comes from sociotechnical systems theory or the other way around, it is the experience of many organizations that employees who are closest to the operational processes must be involved in the organizational design process. When this is not done, the consequences are low employee commitment to the job, incorrect estimates of time, cost, and other factors, omissions of work, and other errors. This happens when the design process involves only technical specialists and senior managers. It is seen most frequently in projects as planning of the project being done by people *other* than those who actually have to carry out the work, and this has been discussed in Chapter 4 on project planning.

Managers often complain that people in their organizations resist change. I don't believe this is true. In my view, people don't resist change, but they *resist being changed*. That is a significant difference. If we make people a part of the change process, then we do not get such resistance.

Another part of this is that *people don't argue with their own data*. I first heard Bob Pike say this. If you make them part of the change process, then they get data first-hand that validates the need for change, so they don't argue with it. Further, the fact that they develop the change process means they do not resist it. These are important facts for managers to remember in trying to bring about improvements in organizational performance.

Assessment of Strengths and Weaknesses of the Organization

One of the practices of project planning that is recommended in Chapter 6 is for the team to conduct a SWOT analysis. The acronym stands for Strengths, Weaknesses, Opportunities, and Threats, with threats and risks being more or less synonymous. The SWOT analysis should always be conducted through a review of objective data whenever possible. When it is dependent only on individual perceptions and general ideas, it is suspect, partly because managers are often too optimistic at the beginning of a project. They tend to minimize weaknesses and threats, and risks are not taken seriously enough. Further, their optimism sometimes causes them to underestimate the number of resources or time required to do the work.

There is, in fact, a *macho* notion prevalent among some managers that one should never admit any weakness—whether organizational or personal. Clearly, failure to admit weakness means that such weakness is not accessible to correction. Chris Argyris (1990) has discussed the processes that prevent organizations from learning from SWOT analyses and postmortems of previous projects. There are two key processes that prevent learning: defensive routines and fancy footwork.

Defensive routines are attempts by members of the organization to avoid embarrassing anyone, so they hide *from* the truth as well as hide the truth from other members of the organization, so that the truth cannot be used to signal the need for change.

Fancy footwork often involves reinterpretation of the data so that it has a favorable meaning. An example of this that is current today (October 1992) is the recent release of unemployment statistics. The unemployment rate dropped by 0.1 percent in September, and this appears to be in part because some people have just given up looking for jobs and so were dropped from the list of unemployed. This would mean, of course, that the situation really is not improved at all. Nevertheless, President Bush appeared on television saying that it was a very positive sign, especially since it was the third consecutive month in which a drop had occurred.

The Need for Joint-Optimization

As has been stated previously, optimum performance of any organization can be achieved only through joint-optimization of the social and technical systems. In the past, we have seen organizations falter because they came under the influence of someone who optimized only the social component. These individuals may have been heavily influenced by the human relations movement of the 1950s and 1960s. Their bias was that what was important in organizations was promoting job satisfaction, good relations, low levels of interpersonal conflict, and so on. In other words, they tried to create a "country club" environment in the workplace. Unfortunately, making people happy does not always correlate with good organizational performance such as quality, productivity, and profitability, and such country-club companies soon found themselves in trouble.

At the opposite extreme is the organization that tries to optimize only the technical component. They invest in state-of-the-art equipment, streamline the work processes, employ statistical process control methodology, and ignore the social system except for the bare minimum requirements. People are allowed to atrophy through lack of training and development. Rewards are dispensed only in the form of money. Conflicts are allowed to reach the boiling level before any attempt is made to resolve them, and then the resolution may take the form of warning the parties involved that it will not be tolerated, rather than trying to get at the root cause so that the conflict is eliminated.

Unfortunately, joint-optimization is much more difficult to achieve than it is to prescribe. Part of the difficulty is in the fact that the systems are not independent. We talk about them separately for convenience, but they are in most cases inter-related and therefore, as has been pointed out above, if a change is made in one, it affects the others. In fact, it may well be the *interaction* effects which are more important than the *first-order* effects.

For example, giving Tom a computer is a first-order effect in the technical component of the system. Whereas Tom previously had to do his calculations manually, he now has new technology, making it possible for him to do the calculations considerably faster than before.

However, he now finds that he is not as free to talk with Charlie as he used to be. Before the computer came along, he and Charlie often met together to work up weekly reports, helping each other with the calculations. Now each has a computer and is expected to perform the calculations individually. This is an *interaction* effect. The change made to the technical system has caused a change in the social system, and this change may have more severe consequences than the first-order effect. The reason: Tom and Charlie both miss their social interaction so much that their morale declines. They begin complaining to their co-workers that the organization is becoming too cold and impersonal for their tastes, and they do so much "rabble-rousing" that soon there is a spread of low morale.

Their manager notices their rabble-rousing and warns them that it must stop. This further confirms that the organization (represented by their boss) has become cold and uncaring about them as human beings. They protest more. The boss finally dismisses them because he cannot live with their *attitudes!*

Another example, which I got from a client organization: They had a piece of equipment that they planned to eliminate because it was considered obsolete. The word got out that this was their plan. Unfortunately, the fellow who had operated the equipment for years saw the handwriting on the wall. If there was no need for his machine, what would happen to him?

He became despondent, believing that he would go out the door with the machine. His performance declined. His boss thought he was trying to retire on the job, and became concerned. Ultimately the employee was forced out of the company—his belief about the company's intentions became a *self-fulfilling prophecy.*

The sad thing is, the company fully intended to move him to another position when they disposed of his machine. He was always considered a valuable employee. Yet no one took the time to tell him what was going on (thereby attending to the social system component). Unfortunately, this is not an isolated incident.

Reactive Versus Proactive Management

Much has been written about the tendency of American management to focus on the short-term rather than the long-term, and to be reactive rather than proactive. This same problem typifies project management in many cases. The project manager becomes so involved in solving today's problems that he or she fails to look ahead, or to interpret the problems as symptoms of a greater "illness" that afflicts the project. This is understandable, when one realizes that we generally respond most strongly to that which is most salient in our experience, and it is the *immediate problems* which clearly are most salient. Tomorrow is "out there" somewhere in never-never land. It is not tangible.

It takes real discipline and the ability to back away from today and look at the "big picture" to get out of this short-term focus. It may even require an outside auditor to help, which is why periodic project audits are recommended as a safeguard against this tendency. (See Chapter 12 on audits.)

Goal Selection and Orientation

There is a tendency for organizations to make choices of goals in an either/or manner. *Either* we can have quality *or* we can have it finished on time, but we can't have both! This sometimes leads to our seeing the customer as the enemy.

I got into a cab at O'Hare airport once and told the driver where I wanted to go. It was in the outskirts of Chicago. As he started to drive away, he said aloud, "The other cabs get to go downtown. Me, I get to go to (my destination)."

I said to him, "Look, if you don't want to take me there, let me out, and I'll get another cab."

"No, it's too late," he said. What he meant was that he had lost his place in the queue and would lose even more if he had to start over.

I must admit to being more than a little steamed. Finally I said to him, "It seems to me you have forgotten who pays your salary. Your customers are not the enemy. If you don't want to take them wherever they want to go, then you should hang a sign on your window saying 'downtown only.'"

He didn't say anything, but I know that he was only thinking about profits. If he went downtown, he had a good chance of picking up someone at a hotel who wanted to go back to the airport. Taking me to a suburban hotel meant that he would probably have to return to the airport empty, thus losing money.

I understand his concern, but for him to dump on me, the customer, was inappropriate.

As has been stated on project missions, it is necessary that a project team understand its primary reason for existence, which is *to satisfy the needs of its customers.* Failure to keep that in mind is certain to lead to failure overall.

As I have said in a previous chapter, the prime *motive* of a business is to make a profit, but its mission must be customer satisfaction. But they are not an either/or choice. They must be achieved simultaneously. The same is true of a project team.

Limits of the "Old Standbys"

As has been stated previously, project organization largely falls into two categories, hierarchical and matrix. These are the old standbys with which everyone is familiar. The problem is, when they are viewed as the only choices, they limit our ability to achieve more optimum solutions. Often we limit ourselves by working from a constraint orientation, which emphasizes what cannot be changed, rather than from an innovative orientation, which looks for what *can* be changed.

We must constantly search for organization forms that solve some of the problems of matrix and hierarchical structure. The most recent trend is cross-function management, which is neither matrix nor hierarchical. In this form of organization, as described by Dimancescu (1992), design-build teams were assembled by Boeing to design the 777 aircraft, which will eventually replace the 747. Nearly 215 teams, of as many as 15 members each, were assembled. Many were co-located, to avoid the problems typically encountered with matrix, in which members are spread out physically so that communication occurs haphazardly, if at all.

This form of organization created a unique situation. Matrix results in a one-person-two-bosses form, which has long been deplored, yet deemed necessary to get complex jobs done. The Boeing organization created a situation of two-bosses-one-hat.

Cross-function management is probably here to stay, for the foreseeable future. Multidisciplinary teams are essential to deal with complex engineering projects, and the old matrix structure has proven to have numerous problems. No doubt cross-function management will as well.

Regarding the Organization Design as "Finished"

To expect that the design of an organization is finished "once and for all," is to limit our new possibilities and to freeze our response capability in the face of changes that make the old design obsolete. It is better to regard design work as a part of regular operations and not a separate front-end activity. This requires setting goals for peo-

Most applications
of sociotechnical systems have
been applied to manufacturing environments.

ple development so that appropriate skills and flexibility are developed as needed to respond to the changing environment.

A Final Caution

While this chapter has suggested that principles from sociotechnical systems design might be applied to project organization, there is a word of caution in order. Most applications of sociotechnical systems principles have been done in manufacturing environments, where jobs tend to be simplified, boring, and nonchallenging. In those cases, cross-training of workers tends to enlarge and enrich their jobs, making them more motivating.

To apply the same ideas to knowledge workers can be risky. For example, the idea of cross-training engineers in project teams was suggested by one practitioner of the art. My response was that this is like trying to teach a brain surgeon to do heart surgery and vice versa. You wind up with two surgeons who are no good at either profession. The reason is simple—it is very nearly impossible to keep up with one's profession now, much less try to learn the skills of another!

Marvin Weisbord (1987), who is regarded by his colleagues as one of the nation's foremost practitioners of organization development, says,

> Anyone who tries to clone this procedure for project management, product development, or planning, quickly discovers that knowledge work happens differently from repetitive production work (a continuing source of irritation between scientists and cost accountants). The flow chart spills out in all directions. People already have multiskilled jobs, with considerable decision latitude (p. 324).

In my opinion, the answer to improving project team performance is in applying the ideas of cross-function management, as discussed above. To that approach, we might apply sociotechnical systems design principles and come up with the project organization of the 21st century.

Chapter 32

How to Achieve Your Goals[*]

Creating Your Ideal Personal Future: The Nature of Personal Premises

The term *personal premises* is used to designate the thought processes in which we engage, including beliefs, assumptions, values, and expectations. Another term that is sometimes used is *model of the*

---

[*] This chapter is based on an exercise developed by John D. Adams, originally published in the *1988 Annual for Group Facilitators*, published by University Associates, San Diego, CA.

world. What do we believe to be true about the world in which we live?

For many of us, our models of the world are not explicit—they are unconscious. Nevertheless, conscious or not, they influence how we interact with the world. In fact, the unconscious nature of our models is what makes them so important: so long as they function at a nonconscious level, they prevent us from gaining control over our lives in order to achieve the results that are important to us.

Self-Fulfilling and Self-Limiting

The self-fulfilling prophecy says that what we hold in our minds (even subconsciously) tends to occur in our lives. Thus, my model of the world tends to be a reality for me. Further, if I dwell on certain things, that experience tends to predominate for me. For example, if I am always concerned about problems, I find more and more problems.

A story from India illustrates this very well. A very wealthy man who was approaching the later years of his life became concerned about his spiritual future. So he went to a renowned guru and asked how to achieve enlightenment. The guru apparently perceived that the man was not really a seeker after truth, but was concerned for his well-being in the afterlife (or the next life).

"It is very simple," the guru told him. "From this moment on, do not think of black birds. Let no thoughts of black birds enter at all, and you will achieve enlightenment."

The man went home, a bit puzzled by this enigmatic instruction. And, no matter how hard he tried, he could not eliminate the thought of black birds from his mind. In fact, they were multiplied by the hundreds, the thousands, perhaps even the millions!

In the case of the self-fulfilling prophecy, as we dwell on problems and find ourselves having more and more of them, that evidence simply reinforces our belief. As the belief becomes more entrenched, it becomes less questionable, thereby limiting our ability to break out of it.

For managers trying to progress in their careers, these same processes can be the limiting factors in their success. For example, we sometimes let "shoulds" govern our lives, such as, "My boss should be more clear in his expectations for me," or "I really should have gotten that job—Charlie didn't deserve it." These "shoulds" can lead to very unrealistic expectations for ourselves and others and result in dysfunctional behaviors on our part. Beliefs that would yield a more positive approach might be, "I need to sit down with my boss and see if we can clarify his expectations for me," or "Let me see why Charlie had an edge over me in getting that job, and then let me see what I need to do to prepare for the next opportunity that comes along."

One of the ways in which we maintain our beliefs is with internal dialogue. We tell ourselves that we were right, that so and so happened exactly as we expected, that nothing will ever change, and so on. We also tend to associate with people who reinforce our beliefs. For example, a person with low self-esteem may associate with someone who is always putting him or her down, thus receiving repeated reinforcement of his or her beliefs about personal worth. All of this serves to maintain the *programming* that we have received in the past, and unless the programming is changed, our future is going to be more of the same things that we have already experienced.

You Can Change Your Programming

We need not be victims of our programming, however. We can change it by choosing new, more desirable approaches and suggesting them to ourselves repeatedly so that new programs replace the old ones. This is not always easy, and requires that you be aware of relationships that maintain the old beliefs and take steps to break out of them. It is also important to note that the *broken-record* approach may not work. You have to be truly introspective, examine your beliefs, and consciously realize that those beliefs can be challenged and changed. Then you may have to start with small steps and gradually work toward the larger objectives.

One approach that can be particularly helpful is to select people with whom you do not normally interact and act *as if* your new beliefs were true, and you will find that they are. Be careful not to try the *as-if* approach with people who know you well. They may well react negatively and undermine your attempts to make the new behavior work. After all, interactions with people tend to become patterns that repeat, and systems tend to resist having those interaction patterns changed.

A Sense of Purpose

It is also helpful to examine your values and develop a sense of purpose in life. What are you really interested in achieving in life? If someone were to write your epitaph, what would you want it to say you did with your life?

How I Stop Myself
From Getting the
Results I Want!

By answering the following questions, you can develop steps to achieve the future you want, rather than the one you may have been programmed for.

1. How I stop myself from getting the results I want:

    Behaviors:

    Beliefs:

2. Transformed behaviors and beliefs. These should be specific, results-focused, positive, and written as if they are already true.

3. My strongest assets in getting the results I want:

4. Ways I currently express these assets:

5. Elements of my ideal world:

6. My purpose in life is to use [my assets] through [the ways I express these assets], in order to foster [the elements of my ideal world].

To help achieve my purpose, I affirm [my transformed beliefs and behaviors].

Glossary

A Glossary of Project Management Terms

Activity The work or effort needed to achieve a result. It consumes time and usually consumes resources.

Activity Description A statement specifying what must be done to achieve a desired result.

Activity-on-Arrow A network diagram showing a sequence of activities, in which each activity is represented by an arrow, with a circle representing an event (see reference) at each end.

Activity-on-Node A network diagram showing a sequence of activities, in which each activity is represented by a box or circle (that is, a *node*), and these are interconnected with arrows to show precedence of work.

Authority The legitimate power given to a person in an organization to use resources to reach

---

an objective and to exercise discipline.

Backward-Pass Calculation Calculations made working backward through a network from the latest event to the beginning event to calculate event late times. A forward-pass calculation (see reference) determines early times.

Calendars The arrangement of normal working days, together with nonworking days, such as holidays and vacations, as well as special work days (overtime periods) used to determine dates on which project work will be completed.

Change Order A document that authorizes a change in some aspect of a project.

Control The practice of monitoring progress against a plan so that corrective steps can be taken when a deviation from plan occurs.

CPM Acronym for Critical Path Method. A network diagramming method that shows the longest series of activities in a project, thereby determining the earliest completion for the project.

Crashing An attempt to reduce activity or total project duration, usually by adding resources.

Critical Path The longest sequential path of activities that are absolutely essential for completion of the project.

Dependency The next task or group of tasks cannot begin until preceding work has been completed, thus the word *dependent* or *dependency*.

Deviation Any variation from planned performance. The deviation can be in terms of schedule, cost, performance, or scope of work. Deviation analysis is the heart of exercising project control.

Dummy Activity A zero-duration element in a network showing a logic linkage. A dummy does not consume time or resources, but simply indicates precedence.

Duration The time it takes to complete an activity.

Earliest Finish The earliest time that an activity can be completed.

Earliest Start The earliest time that an activity can be started.

Estimate A forecast or guess about how long an activity will take, how many resources might be required, or how much it will cost.

Event A point in time. An event is binary. It is either achieved or not, whereas an activity can be partially complete. An event can be the start or finish of an activity.

Feedback Information derived from observation of project activities, which is used to analyze the status of the job and take corrective action if necessary.

Float A measure of how much an activity can be delayed before it begins to impact the project finish date.

Forward-Pass Method The method used to calculate the Earliest Start time for each activity in a network diagram.

Free Float The amount of time that an activity can be delayed without affecting succeeding activities.

Gantt chart A bar chart that indicates the time required to complete each activity in a project. It is named for Henry L. Gantt, who first developed a complete notational system for displaying progress with bar charts.

Hammock activity A single activity that actually represents a group of activities. It "hangs" between two events and is used to report progress on the composite that it represents.

Histogram A vertical bar chart showing (usually) resource allocation levels over time in a project.

***i-j* notation** A system of numbering nodes in an activity-on-arrow network. The *i*-node is always the beginning of an activity, while the *j*-node is always the finish.

Inexcusable Delays Project delays that are attributable to negligence on the part of the contractor, which lead in many cases to penalty payments.

Latest Finish The latest time that an activity can be finished without extending the end date for a project.

Latest Start The latest time that an activity can start without extending the end date for a project.

Learning Curve The time it takes humans to learn an activity well enough to achieve optimum performance can be displayed by curves, which must be factored into estimates of activity durations in order to achieve planned completion dates.

Leveling An attempt to smooth the use of resources, whether people, materials, or equipment, to avoid large peaks and valleys in their usage.

Life Cycle The phases that a project goes through from concept through completion. The nature of the project changes during each phase.

Matrix Organization A method of drawing people from functional departments within an organization for assignment to a project team, but without removing them from their physical location. The project manager in such a structure is said to have *dotted-line* authority over team members.

Milestone An event of special importance, usually representing the completion of a major phase of project work. Reviews are often scheduled at milestones.

Most Likely Time The most realistic time estimate for completing an activity under normal conditions.

Negative Float or Slack A condition in a network in which the *earliest time* for an event is actually later than its *latest time*. This happens when the project has a constrained end date that is earlier than can be achieved, or when an activity uses up its float and is still delayed.

Node An event in a network. Events are always binary—that is, they are achieved or not. An activity, on the other hand, can be partially complete.

PERT Acronym which stands for Program Evaluation and Review Technique. PERT makes use of network diagrams as does CPM, but in addition applies statistics to activities to try to estimate the probabilities of completion of project work.

Pessimistic Time Roughly speaking, this is the *worst-case* time to complete an activity. The term has a more precise meaning, which is defined in the PERT literature.

Phase A major component or segment of a project.

Precedence Diagram An activity-on-node diagram.

Queue Waiting time.

Resource Allocation The assignment of people, equipment, facilities, or materials to a project. Unless adequate resources are provided, project work cannot be completed on schedule, and resource allocation is a significant component of project scheduling.

Resource Pool A group of people who can generally do the same work, so that they can be chosen randomly for assignment to a project.

Risk The possibility that something can go wrong and interfere with the completion of project work.

Scope The magnitude of work that must be done to complete a project.

Subproject A small project within a larger one.

Statement of Work A description of work to be performed.

Time Now The current calendar date from which a network analysis, report, or update is being made.

Time Standard The time allowed for the completion of a task.

Variance Any deviation of project work from what was planned. Variance can be around costs, time, performance, or project scope.

Work Breakdown Structure A method of subdividing work into smaller and smaller increments to permit accurate estimates of durations, ressource requirements, and costs.

Bibliography

Abt Associates Inc., *Applications of Systems Analysis Models: A Survey*, Washington, D.C., Technology Utilization Division, Office of Technology Utilization, National Aeronautics and Space Administration, 1968.

Ackoff, Russell Lincoln, *Redesigning the Future: A Systems Approach to Societal Problems*. New York: John Wiley, 1974.

Ackoff, Russell Lincoln, and Fred E. Emery, *On Purposeful Systems*. Chicago: Aldine/Atherton, 1972.

Adamiecki, Karol, "Harmonygraph," *Przeglad Organizacji* (Polish Journal on Organizational Review), 1931.

Adams, J., S. Barndt, and M. Martin, *Managing by Project Management*, Universal Technology Corp., Dayton, Ohio, 1979.

Adkinson, A.C., and A.H. Bobis, "A Mathematical Basis for the Selection of Research Projects," *IEEE Transactions on Engineering Management*, Jan. 1969.

Adrian, J.J., Estimating, Scheduling, and Cost Control, *Concrete Construction*, February 1988, v33, p.70+.

Agarwal, J.C., "Project Planning at Kennecott," *Research Management*, May 1974.

Agin, Norman, "Optimum Seeking With Branch and Bound," *Management Science*, Dec. 1966.

Alderfer, Clayton P., *Change Processes in Organizations*, New Haven, Conn.: Dept. of Administrative Sciences, Yale Univ., 1971.

Allen, Louis A., *The Professional Manager's Guide* , USA: Louis A. Allen Associates, 1969.

American Society of Tool and Manufacturing Engineers, *Effective Project Management*. Argyle Pub., 1967.

American Society of Civil Engineers (Author and Publisher), *Effective Project Management Techniques*, Argyle Pub., 1973.

Anderson, J., and R. Narasimhan, "Assessing Project Implementation Risk: A Methodical Approach," *Management Science*, June 1979.

Anderson, D.R., D.J. Sweeney, and T.A. Williams, *An Introduction to Management Science*, West Publishing, 1981.

Anderson, S.D., & R.W. Woodhead, *Project Manpower Management: Management Processes in Construction Practice*, New York: Wiley, 1981.

Ang, A. H-S, J. Abdelnour, and A.A. Chaker, "Analysis of Activity Networks Under Uncertainty," *Journal of the Engineering Mechanics Division Proceedings of American Society of Civil Engineers*, Vol. 101, No. EM4, August 1975, pp. 373-387.

Anthony, Robert N., *Planning and Control Systems: A Framework for Analysis*, Boston: Division of Research, Graduate School of Business Administration, Harvard Univ., 1965.

Aptman, L.H., "Project Management: Scheduling Tools and Techniques," *Management Solutions*, October 1986, v31, p.32 (5).

Aptman, L.H., "Project Management: Setting Controls," *Management Solutions,*. November 1986, v31, p.32 (2).

Aptman, L.H., "Project Management: A Process to Manage Change," *Management Solutions*, August 1986, v31, p.30 (5).

Aptman, L.H., "Project Management: Criteria for Good Planning," *Management Solutions*, September 1986, v31, p.22 (5).

Arber, R.P., "Management of Capital Project Development," *Public Works*, February 1992, v123, p.61 (5).

Archibald, Russell D., *Managing High Technology Programs and Projects*, New York: Wiley, 1976.

Archibald, R.D., and R.L. Villoria., *Network Based Management Systems, (PERT/CPM)*, Wiley, 1967.

Argyris, Chris, "Resistance to Rational Management Systems," *Innovation*, Issue 10: (1969), pp. 28-42.

Argyris, Chris, "How Tomorrow's Executives Will Make Decisions," *Think*, Vol. 33, Nov.-Dec. 1967, pp. 18-23.

Argyris, Chris, "Today's Problems with Tomorrow's Organizations," *Journal of Management Studies 4:* (Feb. 1967), pp. 31-55.

ARINC Research Corporation. *Guidebook for Systems Analysis/Cost Effectiveness*. Washington, D.C.: U.S. Dept. of Commerce, National Bureau of Standards; distributed by Clearinghouse for Federal Scientific and Technical Information, 1969.

Arisawa, S., and S.E. Elmaghraby. "Optimum Time-Cost Trade-Offs in GERT Networks," *Management Science,* Vol. 18, No. 11, July 1972, pp. 589-599.

Arnold, K., & M. Stewart, "How to manage production facility projects," *World Oil*, November 1983, v197, p.62 (5).

Arrow, K.J., and L. Hurwicz. *Studies in Resource Allocation Processes*, Cambridge Univ. Press, 1977.

Association for Systems Management. *An Annotated Bibliography for the Systems Professional*, 2nd Ed. Cleveland: Association for Systems Management, 1970.

Atkins, W., "Selecting a project manager," *Journal of Systems Management*, October 1980, v31, p.34 (2).

Augustine, Norman R., *Manual for Projects and Programs*. Boston: Harvard Business School Press, 1989.

Austin, A.L., *Zero-Based Budgeting: Organizational Impact and Effects*, AMACOM, 1977.

Austin, Vincent, *Rural Project Management*. London: Bastford Publications, 1984.

Avots, Ivars. "Why Does Project Management Fail?" *California Management Review* 12 (Fall 1969), pp. 77-82.

Avots, Ivars, "Making Project Management Work: The Right Tools for the Wrong Project Manager," *S,A,M, Advanced Management Journal*, Vol. 40 (Autumn 1975), pp. 20-26.

Awani, Alfred O., *Project Management Techniques*, Princeton, NJ: Petrocelli Books, 1983.

Ayers, R.H., R.M. Walsh, and R.G. Staples, "Project Management by the Critical Path Method," *Research Management*, July 1970.

Bachman, J., et al., "Bases of Supervisory Power: A Comparative Study in Five Organizational Settings," *Control in Organizations*, A. Tannenbaum, Ed., New York: McGraw-Hill, 1968, pp. 229-238.

Bacon, J., *Managing the Budget Function*, National Industrial Conference Board, 1970.

Bainbridge, J. *Health Project Management*, World Health Organization, 1974.

Baker, N.R., and J. Freeland, "Recent Advances in R & D Benefit Measurement and Project Selection Models," *Management Science*, June 1975.

Baker, N.R., and W.H. Pound, "R & D Project Selection: Where We Stand," *IEEE Transactions on Engineering Management*, Dec. 1964.

Baker, Frank, Ed. *Organizational Systems; General Systems Approaches to Complex Organizations.* Homewood, IL: R.D. Irwin Series in Management and the Behavioral Sciences, 1973.

Baker, N.R., "R & D Project Selection Models: An Assessment," *IEEE Transactions on Engineering Management,* Nov. 1974.

Baker, B.N., and R.L. Eris, *An Introduction to PERT-CPM,* Irwin, 1964.

Balachandra, R., and A.J. Raelin, "How to Decide When to Abandon a Project," *Research Management,* July 1980.

Balas, E., "Project Scheduling with Resource Constraints," *Applications of Mathematical Programming Techniques,* Carnegie-Mellon Univ., 1970.

Barnes, N.M.L., "Cost Modelling—An Integrated Approach to Planning and Cost Control," *American Association of Chemical Engineers Transactions,* March 1977.

Barrett, J.E., "How to Manage a Crash Project," *Management Review,* September 1974, v63, p.5 (8).

Barrie, Donald S., & Boyd C. Paulson, Jr., *Professional Construction Management* (2nd Ed,). New York: McGraw-Hill, 1984.

Bartizal, J.R., *Budget Principles and Procedures,* New York: Prentice Hall, 1940.

Battersby, A., *Network Analysis for Planning and Scheduling,* New York: St. Martins Press, Inc., 1964, Chapter 3.

Baumgartner, John Stanley, *Project Management,* Homewood, IL: R. D. Irwin, 1963.

Becker, R.H., "Project Selection for Research, Product Development and Process Development," *Research Management,* Sept. 1980.

Beckett, John A., *Management Dynamics: The New Synthesis,* New York: McGraw-Hill, 1971.

Begley, Francis D., *Project Management for Construction Superintendents* (2nd Edition), Saunders of Toronto, Boston: Cahners Books, 1974.

Benne, K.D., and M. Birnbaum, "Principles of Changing" in Bennis, W.G., et al., *The Planning of Change*, New York: Holt, Rinehart, and Winston, 1969.

Bennigson, Lawrence, *Project Management*, McGraw-Hill, 1970.

Bennigson, Lawrence, "The Team Approach to Project Management," *Management Review 61*, (Jan. 1972), pp. 48-52.

Benningson, L.A., "The Strategy of Running Temporary Projects," *Innovation*, Sept. 1971.

Bennington, L.A., *TREND—New Management Information From Networks*, Sandoz Co., Reprint 1974.

Bennis, Warren G., "The Coming Death of Bureaucracy," *Think,* 32, (Nov.-Dec. 1966), pp. 30-35.

Bensen, L.A., and R.F. Sewall, "Dynamic Crashing Keeps Projects Moving," *Computer Decisions*, Feb. 1972, pp. 14-18.

Bent, J.A., "Project Control Concepts," *Project Management Proceedings*, 1979.

Bent, James A., and Albert Thumann, *Project Management for Engineering and Construction*, Lilburn, Ga.: Fairmont Press, Englewood Cliffs, NJ.: Distributed by Prentice Hall, 1989.

Benton, John Breen, *Managing the Organizational Decision Process*, Lexington, Mass.: Lexington Books, 1973.

Berger, W.C., "What a Chief Executive Should Know About Major

Project Management," *Price Waterhouse Review*, Summer/Autumn 1972.

Berger, Seymour, *Estimating and Project Management for Small Construction Firms*, New York: Van Nostrand Reinhold Company, 1977.

Berkwitt, G.J., "Management Rediscovers CPM," *Dun's Review*, May 1971, Dun & Bradstreet Pub. Corp.

Berlinski, David J., "On Systems Analysis: An Essay Concerning the Limitations of Some Mathematical Methods in the Social, Political, and Biological Sciences," Cambridge, Mass: M.I.T. Press, 1976.

Berman, E.B., "Resource Allocation in a PERT Network Under Continuous Activity Time-Cost Functions," *Management Science*, July 1964.

Berrien, F. Kenneth, *General and Social Systems*, New Brunswick, NJ: Rutgers Univ. Press, 1968.

Bertalanffy, Ludwig von, *General Systems Theory; Foundations, Development, Applications*, New York: G. Braziller, 1972.

Bhasin, R., "The Successful Project," *Pulp and Paper*. March 1989, v63, p.167.

Bildson, R.A., and J.R. Gillespie, "Critical Path Planning PERT Integration," *Operations Research*, Nov.—Dec. 1962.

Bingham, John E., and G.W.P. Davies, *A Handbook of Systems Analysis*, London: Macmillan, 1972, 1974. Distributed in North America by Halsted Press, a division of John Wiley, New York and Toronto.

Blacksburg: Virginia Polytechnical Institute and State University (Publisher Only), *Proposal Preparation and Project Management: Manual of Procedures*, 1977.

Blake, R.R. and J.S. Mouton, *The Managerial Grid*, Houston: Gulf Publishing, 1964.

Blanchard, Frederick L., *Engineering Project Management*, New York: M. Dekker, 1990.

Blankstein, Charles Sidney, *The Base Level Development Assistance Project: A Managerial Perspective*, M.I.T., ms., 1972, Cambridge, Mas. Thesis, M.S.

Block, Ellery B., "Accomplishment/Cost: Better Project Control," *Harvard Business Review*, 49, (May 1971), pp. 110-124.

Blystone, E.E., and R.G. Odum, "A Case Study of CPM in a Manufacturing Situation," *Journal of Industrial Engineering*, Nov.-Dec. 1964.

Bobrowski, T.M., "A Basic Philosophy of Project Management," *Journal of Systems Management*, May-June, 1974.

Boder, A., "Getting Projects Back on Schedule," *Machine Design*, September 10, 1987, v49, p142 (2).

Boulding, Kenneth, "General Systems Theory—the Skeleton of Science," *Management Science*, April 1956, pp. 197-208.

Bowenkamp, R.D., and B.H. Kleiner, "How to Be a Successful Project Manager," *Industrial Management and Data Systems*, March/April 1987, p.3 (5).

Bowman, R.R., "An Analysis of Project Management Concepts in the Missile/Space Industry," MBA Thesis, Utah State Univ., 1967.

Boyatzis, R.E., "Leadership: The Effective Use of Power," *Management of Personnel Quarterly*, Graduate School of Business Administration, University of Michigan (Fall 1971), pp. 21-25. Reprinted in Richards, Max D., and William A. Nielander, *Readings in Management*, 4th Ed., (Cincinnati: Southwestern Pub. Co., 1974), pp. 623-629.

Boyatzis, R.E., "Building Efficacy: An Effective Use of Managerial Power," *Industrial Management Review*, Vol. 11, No. 1 (Fall 1969), pp. 65-75.

Brandon, Dick H., and Max Gray, *Project Control Standards*, Princeton, NJ: Brandon/Systems Press, 1970.

Brennan, J., *Applications of Critical Path Techniques*, American Elsevier, 1968.

Briggs, G.R., *The Theory and Practice of Management Control*, American Management Association, 1970.

Brooks, F.P., *The Mythical Man-month: Essays on Software Engineering*, Reading, MA: Addison, 1975.

Brooks, N.A.L., "Managing Systems Development," *Bank Management*, July 1991, v67, p. 44 (4).

Brown, R., and J.D. Suver, "Where Does Zero-Base Budgeting Work?" *Harvard Business Review*, Dec. 1977.

Buell, C.D., "When to Terminate a Research and Development Project," *Resource Management,*, 10, 275-284 (July 1967).

Bunge, W.R., *Managerial Budgeting for Profit Improvement*, McGraw-Hill, 1968.

Burgess, A.R., and J.B. Killebrew, "Variation in Activity Level on a Cyclic Arrow Diagram," *Journal of Industrial Engineering*, March-April 1962.

Burgess, Roger A, and G. White, *Building Production and Project Management*, Lancaster, England; New York: Construction Press, 1979.

Burke, R.J., "Methods of Resolving Interpersonal Conflict," *Personnel Administration*, July-Aug. 1969, pp. 48-55.

Burke, R.J., "Methods of Managing Superior-Subordinate Conflict," *Canadian Journal of Behavioral Science*, 2,2: 1970, pp. 124-135

.Burke, W.W. and H.A. Hornstein, *The Social Technology of Organization Development*, Fairfax, Va.: NTL Learning Resource Corp., 1972.

Burkhead, J., *Budgeting and Planning*, General Learning Press, 1971.

Burrill, Claude W., Leon W. Ellsworth, *Modern Project Management: Foundations for Quality and Productivity*, Tenafly, NJ: Burrill-Ellsworth Associates, 1980.

Burstein, David, and Frank Stasiowski, *Project Management for the Design Professional*, New York: Whitney Library of Design 1982.

Burstein, David. *Project Management for the Design Professional: A Handbook for Architects, Engineers, and Interior Designers* (rev. ed.), New York: Whitney Library of Design, 1991.

Burt, David N., "Getting the Right Price With the Right Contract," *Management Review*, May 1976, pp. 24-34.

Busch, Dennis H., *The New Critical Path Method*, Chicago: Probus, 1991.

Butler, Arthur G., Jr., "Project Management: A Study in Organizational Conflict," *Academy of Management Journal*, 16, March 1973, pp. 84-101.

Butler, D., and N. Miller, "Power to Reward and Punish in Social Interaction," *Journal of Experimental Social Psychology*, Vol. 1, No. 4, (1965), pp. 311-322.

Cammann, C., and D.A. Nadier, "Fit Control Systems to Your Management Style," *Harvard Business Review.*, Jan.-Feb. 1976.

Carruthers, J.A., and A. Battersby, "Advances in Critical Path Methods," *Operational Research Quarterly*, Dec. 1966.

Casley, Dennis J., and Krishna Kumar, *Project Monitoring and Evaluation in Agriculture*. Baltimore: published for the World Bank by Johns Hopkins University Press, 1987.

Caspe, M.S., "Monitoring People to Perform on Design and Construction Projects," *Project Management Quarterly*, Dec. 1979.

Cerullo, M.J., "Determining Post-Implementation Audit Success," *Journal of Systems Management*, March 1979.

Cestin, A.A., "What Makes Large Projects Go Wrong," *Project Management Institute Quarterly*, March 1980.

Chapman, Richard L., *Project Management in NASA—the System and the Men*. Washington: Scientific and Technical Information Office, National Aeronautics and Space Administration; for sale by the Superintendent of Documents, U.S. Government Printing Office, 1973.

Charnes, A., and W.W. Cooper, "A Network Interpretation and a Directed Subdual Algorithm for Critical Path Scheduling," *Journal of Industrial Engineering*, Vol. 13, No. 4(1962), pp. 213-218.

Chen, Gordon, K., and Eugene E. Kaczka, *Operations and Systems Analysis; A Simulation Approach*, Boston: Allyn and Bacon, 1974.

Churchman, Charles West, *The Systems Approach*, New York: Dell Publishing Co., 1968.

Cicero, John P., and David L. Wilemon, "Project Authority: A Multidimensional View," *IEEE Transactions on Engineering Management*, EM-17 (May 1970), pp. 52-57.

Clark, C. G., D.G. Malcom, J.H. Rosenbloom, and W. Fazar, "Applications of a Technique for Research and Development Program Management," *Operations Research*, Sept.-Oct. 1959.

Clark, E., "The Optimum Allocation of Resources Among the Activities of a Network," *Journal of Industrial Engineering*, Jan.-Feb. 1961.

Clark, C.E., "The PERT Model for the Distribution of an Activity Time," *Operations Research*, Vol. 10, No. 3, May-June 1962, pp. 405-406.

Clark, P., "A Profitability Project Selection Method," *Research Management*, Nov. 1977.

Clark, C.E., "The Greatest of a Finite Set of Random Variables," *Operations Research*, Vol. 9, No. 2, March-April 1961, pp. 145-162.

Clarke, W., "The Requisites for a Project Management System," *Project Management Institute Proceedings*, 1979.

Clayton, R., "A Convergent Approach to R & D Planning and Project Selection," *Research Management*, Sept. 1971.

Cleland, David I., and William R. King, (Eds). *Project Management Handbook*. New York: Van Nostrand Reinhold Company, 1983.

Cleland, David I., "Why Project Management?" *Business Horizons*, 7 (Winter 1964), pp. 81-88.

Cleland, David I., *Systems Analysis and Project Management*, New York: McGraw-Hill, 1975.

Cleland, David I., "Organizational Dynamics of Project Management," *IEEE Transactions on Engineering Management*, EM-13 (Dec. 1966), 201-5.

Cleland, David I., and William R. King, *Management: A Systems Approach*, New York: McGraw-Hill, 1972.

Cleland, David I., "Defining a Project Management System," *Project Management Quarterly*, Vol. 8, No. 4 (1977), pp. 37-40.

Cleland, David I., *Systems, Organizations, Analysis, Management; A Book of Readings*, New York: McGraw-Hill, 1969.

Cleland, David I., "The Deliberate Conflict," *Business Horizons*, Vol. 11, No. 1 (1968), pp. 78-80.

Cleland, David I., "Project Management in Industry: An Assessment," *Project Management Quarterly*, Vol. 5, No. 2 and 3 (1974), pp. 19-21.

Clifton, D.S., Jr., and D.E. Fyffe, *Project Feasibility Analysis: A Guide to Profitable Ventures*. New York: Wiley, 1977.

Clough, Richard Hudson, and Glenn A Sears, *Construction Project Management* (3rd Ed.), New York: Wiley, 1991.

Clough, Richard Hudson, and Glenn A. Sears, *Construction Project Management* (2nd Edition), New York: Wiley, 1979.

Clough, Richard Hudson, *Construction Project Management*, New York: Wiley-Interscience, 1972.

Cochran, M., E.B. Pyle, III, L.C. Greene, H.A. Clymer, and A.D. Bender, "Investment Model for R & D Project Evaluation and Selection," *IEEE Transactions on Engineering Management*, Aug. 1971.

Cukierman, A., and Z.F. Shiffer, "Contracting for Optimal Delivery Time in Long-Term Projects," *The Bell Journal of Economics*, Vol. 7, No. 1, Spring 1976, pp. 132-149.

Cook, Desmond Lawrence, *Educational Project Management*, Columbus Ohio: Merrill, 1971.

Cooper, M.J., "Evaluation System for Project Selection," *Research Management*, July 1979.

Corwin, B.D., "Multiple R and D Project Scheduling With Limited Resources," *Technical Memorandum No, 122*, Dept. of Operations Research, Case Western Reserve Univ., 1968.

Couger, J. Daniel, and Robert W. Knapp, Eds., *System Analysis Techniques*, New York: John Wiley, 1974.

Coxon, R., "How Strategy Can Make Major Projects Prosper," *Management Today*, April 1983, p.30 (+).

Crandall, K.C., "Probabilistic Time Scheduling," *Journal of the Construction Division, Proceedings of American Society of Civil Engineers*, Vol. 102, No. CO3, Sept. 1976, pp. 415-423.

Crandall, K.C., "Analysis of Schedule Simulations," *Journal of the Construction Division, Proceedings of American Society of Civil Engineers*, Vol. 103, No. CO3, Sept. 1977, pp. 387-394.

Crandall, K.C., "Project Planning With Precedence Lead/Lag Factors," *Project Management Quarterly*, Vol. 4, N. 3, Sept. 1973, pp. 18-27.

Croft, F.M., "Putting a Price Tag on PERT Activities," *Journal of Industrial Engineering*, July 1966.

Crowston, W., and G.L. Thompson, "Decision CPM: A Method for Simultaneous Planning, Scheduling, and Control of Projects," *Operations Research*, May–June 1967.

Crowston, Wallace B., "Models for Project Management," *Sloan Management Review*, 12 (Spring 1971), pp. 25–42.

Cullingford, G. and J.D.C.A. Prideaux, "A Variational Study of Optimal Resource Profiles," *Management Science*, 19 (May 1973), pp. 1067–81.

Currid, C., "There's a Lot More to Managing Your Projects Than Paper Pushing," *Information World*, January 20, 1992, v14, p.58 (1).

Dahl, R., "The Concept of Power," *Behavioral Science*, Vol. 2 (July 1957), pp. 201–215.

Darnell, H., "Towards Total Project Management," *Management Today*, August 1982, p.60 (4).

Darnell, H., and M.W. Dale, "Total Project Management: An Integrated Approach to the Management of Capital Investment Projects," *Proceedings of the Institution of Mechanical Engineers, Part D, Transport Engineering*, December 1982, v196, p.337 (10).

Datz, M., "Project Management: Develop Project Scope Early," *Hydrocarbon Process*, September 1981, v60, p.161 (16).

Datz, Marvin A. and L.R. Wilby, "What Is Good Project Management?" *Project Management Quarterly*, Vol. 8, No. 1 (March 1977).

Davies, E.M., "An Experimental Investigation of Resource Allocation in Multiactivity Projects," *Operational Research Quarterly*, Dec. 1973.

Davis, E.W., and J.H. Patterson, "Resource-Based Project Scheduling: Which Rules Perform Best?" *Project Management Quarterly*, Sept. 1976.

Davis, E.W., "Project Scheduling Under Resource Constraints: Historical Review and Categorization of Procedures," *AIIE Transactions*. Dec. 1973.

Davis, E.W. (Ed.), *Project Management: Techniques, Applications, and Managerial Issues*. (2nd ed.). Norcross, Georgia: Industrial Engineering and Management Press Institute of Industrial Engineers, 1985.

Davis, E.W., "CPM Use in Top 400 Construction Firms," *Journal of the Construction Division*, ASCE, Vol. 100, No. CO1, Proc. Paper 10395, March 1974, pp. 39-49.

Davis, E.W., "Networks: Resource Allocation," *Industrial Engineering*, Vol. 6, No. 4, April 1974, pp. 117-120.

Davis, Keith, "The Role of Project Management in Scientific Manufacturing," *IRE Transactions on Engineering Management*, Vol. 9, No. 3 (1962).

Davis, Edward W., and George E. Heidorn, "An Algorithm for Optimal Project Scheduling Under Multiple Resource Constraints," *Management Science*, August 1971.

Davis, Keith, "The Role of Project Management in Scientific Manufacturing," *Arizona Business Bulletin*, 9 (May 1962), pp. 1-8.

Davis, Gordon, President, DDR International, Atlanta, Ga., personal interview, November, 1981.

Davis, S.M., and P.R. Lawrence, *Matrix*, Reading, Mass.: Addison-Wesley, 1977.

Davis, S., "An Organic Problem-Solving Method of Organizational Change," *Journal of Applied Behavioral Science* (Jan. 1967), pp. 3-21.

Davis, E.W., and J.H. Patterson, "A Comparison of Heuristic and Optimum Solutions in Resource-Constrained Project Scheduling," *Management Science*, April 1975.

Davis, Stanley, "Two Models of Organization: Unity of Command Versus Balance of Power," *Sloan Management Review*, Fall 1974, pp. 29-40.

Davis, E.W., "Project Network Summary Measures and Constrained Resource Scheduling," *AIIE Transactions*, June 1975.

De Greene, Ed., *Systems Psychology*, New York: McGraw-Hill, 1970.

De Greene, Kenyon Brenton, *Sociotechnical Systems; Factors in Analysis, Design, and Management.* Englewood Cliffs, NJ: Prentice Hall, 1973.

Dean, B.V., *Evaluating, Selecting and Controlling R & D Projects*, American Management Assoc., 1968.

Dean, B.V., *Evaluating, Selecting, & Controlling R & D Projects*, American Management Assoc., Inc., New York, 1968.

Dean, B.V., D.R. Denzler, and J.J. Watkins, "Multi-Project Staff Schedulin With Variable Resource Constraints," *IEEE Transactions on Engineering Management*. February 1992, v39 p.59 (14).

Dean, B.V., S.J. Mantel, Jr., and L.A. Roepcke, "Research Project Cost Distrbutions and Budget Forecasting," *IEEE Transactions on Engineering Management*, Nov. 1969.

Dean, B.V., and L.A. Roepcke, "Cost Effectiveness in R & D Organizatioal Resource Allocation," *IEEE Transactions on Engineering Management*, Nov. 1969.

Deardon, J., *Cost and Budget Analysis*, Englewood Cliffs, NJ: Prentice Hall, 1962.

DeCotiis, T.A., and L. Dyer, "Defining and Measuring Project Performnce," *Research Management*, Jan. 1979.

DeCoster, D.T., "PERT/Cost-The Challenge," *Management Services*, May-June 1964.

Dekom, A.K., "Project Evaluation: Bringing in the Sheaves," *Journal of Systes Management*, December 1991, v42, p.13 (4).

Delbecq, Andre L., and Alan C. Filley, *Program and Project Management n a Matrix Organization: A Case Study*, Madison, WI: Univ. of Wisconsin, Bureau of Business Research and Service, 1974.

Delbecq, Andre L., Gremont A. Schull, Alan C. Filley, and Andrew J. Grims, *Matrix Organization: A Conceptual Guide to Organizational Variation*, Wisconsin Business Papers No. 2, Madison: Univ. of Wisconsin, Bureau of Business Research and Service, 1974.

Dibble, E.T., and Waino Soujanen, "Project Management in a Crisis Economy," *Infosystems-Spectrum*, Vol. 23 (Jan. 1976), pp. 44-46.

Dinsmore, Paul C., *Human Factors in Project Management*. New York, American Management Association, 1984.

Doering, Robert D., "An Approach Toward Improving the Creative Output of Scientific Task Teams," *IEEE Transactions on Engineering Management*, EM-20 (Feb. 1973), pp. 29–31.

Domke, W.V., "Project Planning, Scheduling, and Control," *Buildings*, June 1977, v71, p.68 (3).

Dooley, A.R., "Interpretations of PERT (Keeping Informed)," *Harvard Business Rev.*, March-April 1964.

Dragun, George, "An Accountant's Guide to Project Management," *CMA—The Management Accounting Magazine*, October 1989, v63, p.35 (4).

Drexl, A., "Scheduling of Project Networks by Job Assignment," *Management Science*, December 1991, v37, p.1590 (13).

Drigani, Fulvio, *Computerized Project Control*, New York: Dekker, 1989.

Earle, V.H., "Once Upon a Matrix: A Hindsight on Participation," *Optimum*, 4, No. 2, 1973, pp. 28-36.

Eirich, Peter Lee, "An Information System Design Analysis for a Research Organization," Cambridge, Mass., M.S. Thesis, M.I.T., 1974.

Eisner, H., "A Generalized Network Approach to the Planning and Scheduling of a Research Project," *Operations Research*, Vol. 10, No. 1, 1962, pp. 115-125.

Eiteman, J.W., *Graphic Budgets*, Masterco Press, 1949.

Elliott, D.P., "Paper and Cost Control," *Project Management Proceedings*, 1979.

Elmaghraby, S.E., "An Algebra for the Analysis of Generalized Activity Networks," *Management Science*, Vol. 10, No. 3, 1964, pp. 494-514.

Elmaghraby, S.E., *Activity Networks: Project Planning and Control by Network Models*, New York: John Wiley & Sons, 1977.

Elmaghraby, S.E., and P.H. Pulat, "Optimal Project Compression With Due-Dated Events," *Naval Research Logistics Quarterly*, Vol. 26, No. 2, June 1979, pp. 331-348.

Emery, J.C., *Organizational Planning and Control Systems*, New York: Macmillan, 1969.

Emery, F.E., *Systems Thinking: Selected Readings*, New York: Penguin Education, 1974.

Emshoff, James R., *Analysis of Behavioral Systems*, New York: Macmillan, 1971.

Enrick, N.L., "Value Analysis for Priority Setting and Resource Allocation," *Industrial Management*, Sept.-Oct. 1980.

European Industrial Research Management Association, "Top-Down and Bottom-Up Approaches to Project Selection," *Research Management*, March 1978.

European Conference on the Management of Large Space Programs, Paris, 1970, London, New York: Gordon and Breach Science Publishers, 1971.

Evan, W.M., "Conflict and Performance in R & D Organization," *Industrial Management Review*, Vol. 7, 1965, pp. 37–45.

Evan, W.M., "Superior-Subordinate Conflict in Research Organizations," *Administrative Science Quarterly*, July 1965, pp. 52–64.

Evarts, H.E., *Introduction to PERT*, Allyn & Bacon, 1964.

Exton, William, *The Age of Systems: The Human Dilemma*, New York: American Management Association, 1972.

Federal Power Commission exhibit EP–237, "Risks Analysis of the Arctic Gas Pipeline Project Construction Schedule," Vol. 167, Federal Power Commission, 1976.

Fendley, Larry G., "Toward the Development of a Complete Multi-Project Scheduling System," *Journal of Industrial Engineering*, October 1968, unpublished Ph.D. thesis, Arizona State University, 1966, same title.

Fiore, Michael V., "Out of the Frying Pan Into the Matrix," *Personnel Administration 33*, No. 3, 1970, pp. 4-7.

Fisher, Gene Harvey, *Cost Considerations in Systems Analysis*, New York: American Elsevier, 1971.

Fisk, Edward R., *Construction Project Administration*, New York: Wiley, 1978.

FitzGerald, John M. and Ardra F., *Fundamentals of Systems Analysis*, New York: Wiley, 1971.

Flaks, Marvin, and Russell D. Archibald, "The EE's Guide to Project Management," *Electronic Engineer*, 27, April 1968, pp. 28+, May, pp. 20+, June, pp. 27-32, July, pp. 33-34+, Aug., pp. 33+.

Fleming, Q.W., J.W. Bronn, and G.C. Humphreys, *Project and Production Scheduling*, Chicago: Probus, 1987.

Fleming, Q.W., and Fleming, Q.J., *Subcontract Project Management and Control: Progress Payments*, Chicago: Probus, 1992.

Fleming, Q.W., *Cost/Schedule Control Systems Criteria*, Chicago: Probus, no date.

Fondahl, J.W., "A Noncomputer Approach to the Critical Path Method for the Construction Industry," Dept. of Civil Engineering, Stanford University, Stanford, Calif., 1st Ed., 1961, 2nd Ed., 1962.

Forrester, Jay W., "A New Corporate Design," *Industrial Management Review* 7, Fall 1965, pp. 5-17.

Fox, S., "Organizing Projects," *Supervisory Management*, April 1982, v27 p.34 (4).

Foxhall, William B., *Professional Construction Management and Project Administration* (2nd Ed,). New York: Architectural Record Books, 1976.

Frame, J. Davidson, *Managing Projects in Organizations: How to Make Best Use of Time, Techniques, and People* (1st edition), San Francisco: Jossey-Bass, 1988.

Frankel, Ernest G., *Project Management in Engineering Services and Development*, London; Boston: Butterworths, 1990.

Frankwicz, M.J., "Study of Project Management Techniques," *Journal of Systems Management*, October 1973, v24, p.18 (5).

Frankwicz, Michael J., "A Study of Project Management Techniques," *Journal of Systems Management*, 24, Oct. 1973, pp. 18-22.

Frazier, Haugg, and Thackery, "Developing a Project Management Package," *Journal of Systems Management*, Dec. 1976.

Freeman, R.J., "A Generalized PERT," *Operations Research*, Vol. 8, No. 2, 1960, p. 281.

French, J.R., Jr., and B. Raven, "The Bases of Social Power," in *Studies in social Power*, D. Cartwright, Ed., Ann Arbor, Mich.: Research Center for Group Dynamics, 1959, pp. 150–165.

Fried, Louis, "Don't Smother Your Project in People," *Management Advisor*, 9 (March 1972), pp. 46-49.

Friend, Fred L., "Be a More Effective Program Manager," *Journal of Systems Management*, Vol. 27, (Feb. 1976), pp. 6–9.

Fulkerson, D.R., "Expected Critical Path Lengths in PERT Networks," *Operations Research*, Vol. 10, N. 6, Nov.–Dec. 1962, pp. 808–817.

Fuller, D., "How to Manage Engineering Projects," *Machine Design*, June 23, 1983, v55, p.55 (3); July 7, 1983, v55, p.61 (4); July 21, 1983, v55, p.57 (5); August 25, 1983, v55, p.53(6) and continued in September 8, 1983, v55, p.55 (4).

Gaddis, P.O., "The Project Manager," *Harvard Business Review*, Sept-Oct., 1962.

Galbraith, Jay R., "Matrix Organization Designs—How to Combine Functional and Project Forms," *Business Horizons*, Feb. 1971.

Galbraith, Jay, *Designing Complex Organizations*, Reading, Mass.: Addison-Wesley Publishing Co., 1973.

Gautschi, T.F., "Project Management," *Design News*, October 19, 1981, v37, p.177; November 2, 1981, v37, p.231; November 16, 1981, v37, p.134 (2); December 7, 1981, v37, p.233; December 21, 1981, v37, p.87; January 4, 1982, v38, p.170; January 18, 1982, v38, p.163; February 1, 1982, v38, p.191; February 15, 1982, v38, p.147; March 1, 1982, v38, p.108 (2); March 15, 1982, v38, p.383; April 5, 1982, v38, p.115; April 19, 1982, v38, p.207.

Gee, R.E., "A Survey of Current Project Selection Practices," *Research Management*, Sept. 1971.

Geisler, M.A., "How to Plan for Management in New Systems," *Harvard Business Review*, Sept.-Oct. 1962.

Gemmill, G.R., and H. Thamhain, "Influence Styles of Project Managers: Some Project Performance Correlates," *Academy of Management Journal*, Vol. 17, No. 2 (June 1974), pp. 216-224.

Gemmill, G.R., "Managerial Role Mapping," *The Management Personnel Quarterly*, Vol. 8, No. 3, (Fall 1969), pp. 13-19.

Gemmill, G.R., and H. Thamhain, "The Effectiveness of Different Power Styles of Project Managers in Gaining Project Support," *IEEE Transactions on Engineering Management EM-20* May 1973, pp. 38-44.

Gemmill, G.R., "Interpersonal Power in Temporary Management Systems," *Journal of Management Studies*, Oct. 1971.

Gemmill, G.R., and H. Thamhain, "The Power Styles of Project Managers: Some Efficiency Correlates," *20th Annual JEMC, Managing for Improved Engineering Effectiveness* (Atlanta, GA, Oct. 30-31, 1972) pp. 89-96.

Gemmill, G.R., and H. Thamhain, "Project Performance as a Function of the Leadership Styles of Project Managers: Results of a Field Study," *Convention Digest, 4th Annual Meeting of the Project Management Institute*, Philadelphia, October 18-21, 1972.

Gemmill, G.R., and David L. Wilemon, "The Power Spectrum in Project Management," *Sloan Management Review 12*, Fall 1970, pp. 15-25.

Gemmill, G.R. and David L. Wilemon, "The Product Manager as an Influence Agent," *Journal of Marketing*, Vol. 36 (Jan. 1972), pp. 26-31.

Gibson, James L., Ed. *Readings in Organizations: Structure, Processes, Behavior*, Dallas: Business Publication, 1973.

Gildersleeve, Thomas Robert, *Data Processing Project Management*, Van Nostrand Reinhold Company, 1974.

Gill, P.G., *Systems Management Techniques for Builders and Contractors*, New York: McGraw-Hill, 1968.

Goggin, William C., "How the Multidimensional Structure Works at Dow Corning," *Harvard Business Review*, Jan.-Feb. 1974, pp. 54-65.

Golabi, K., G.W. Kirkwood, and A. Sicherman, "Selecting a Portfolio of Solar Energy Projects Using Multi-Attribute Preference Theory," *Management Science*, Feb. 1981.

Golfarb, N., and W.K. Kaiser, *Gantt Charts and Statistical Quality Control*, Hofstra Univ. Press, 1964.

Goodman, Richard A., "Organizational Preference in Research and Development," *Human Relations 23*, 1970, pp. 279-298.

Goodman, Richard A., "Ambiguous Authority Definitions in Project Management," *Academy of Management Journal 10*, Dec. 1967, pp. 395-408.

Gordon, J.H., "Heuristic Methods in Resource Allocation," *Proceedings of the 4th Internet Conference*, Paris 1974.

Gorenstein, S., "An Algorithm for Project (Job) Sequencing With Resource Constraints," *Operations Research*, July-Aug. 1972.

Gossom, W. J., *Control of Projects, Purchasing, and Materials*, Tulsa, OK: Penn Well Books, 1983.

Gray, Clifford F., *Essentials of Project Management*, Princeton, NJ: PBI, 1981.

Gray, C.F., *Essentials of Project Management*, Petrocelli Books, 1981.

Gretes, Frances C., *Architectural Project Management*, Monticello, IL: Vance Bibliographies, 1984.

Grimes, A., S. Klein, and F. Shull, "Matrix Model: A Selective Empirical Test," *Academy of Management Journal*, Vol. 15, No. 1 (Mar. 1972), pp. 9-31.

Grinnell, S.K., and H.P. Apple, "When Two Bosses are Better Than One," *Machine Design*, 9 (Jan. 1975), pp. 84-87.

Gross, Paul F., *Systems Analysis and Design for Management*, New York: Dun-Donnelley, 1976.

Gullet, C. Ray, "Personnel Management in the Project Organization," *Personnel Administration and Public Personnel Review*, 1 Nov. 1972, pp. 17-22.

Gunderman, J.R., and F.W. McMurry, "Making Project Management Effective," *Journal of Systems Management*, Feb. 1975, v26 p.7 (5).

Gunz, H.P., and A. Pearson, "How to Manage Control Conflicts in Project Based Organizations," *Research Management*, March 1979.

Hajek, Victor G., *Management of Engineering Projects* (3rd Ed.), New York: McGraw-Hill, 1984.

Hajek, Victor G., *Management of Engineering Projects* (2nd Ed.), New York: McGraw-Hill, 1977.

Hajek, S.V.G., *Management of Engineering Projects*, New York: McGraw-Hill, 1977.

Hall, H. Lawrence, "Management: A Continuum of Styles," *S,A,M, Advanced Management Journal* 33, Jan. 1968, pp. 68–74.

Hall, D.M., *Management of Human Systems*, Cleveland, OH: Association for Systems Management, 1971.

Hansen, J.J., "The Case of the Precarious Program," *Harvard Business Review*, Jan.–Feb, 1968.

Hastings, N.A.J., "On Resource Allocation in Networks," *Operational Research Quarterly*, June 1972.

Haviland, David S., *Managing Architectural Projects: The Project Management Manual*, Washington, DC: American Institute of Architects, 1984.

Haviland, David S., *Managing Architectural Projects: The Effective Project Manager*. Washington, D.C.: American Institute of Architects, 1981.

Health Research, Center for. "Health Research: The Systems Approach," New York: Springer, 1976.

Helin, A.F., and W.E. Souder, "Experimental Test on a Q-Sort Procedure for Prioritizing R & R Projects," *IEEE Transactions*, Nov. 1974.

Hellriegel, Don, and John W. Slocum, Jr., "Organizational Design: a Contingency Approach," *Business Horizons*, Vol. 16, No. 2, (April 1, 1973), pp.59-68. Reprinted in Richards, Max D. and William A. Nielander, *Readings in Management*, 4th ed. (Cincinnati: Southwestern, 1974), pp. 516-527.

Hendrickson, Chris, *Project Management for Construction: Fundamental Concepts for Owners, Engineers, Architects, and Builders*, Englewood Cliffs, NJ: Prentice Hall, 1989.

Herbert, J.E., III, "Project Management With PROJECT/GERT," ORSA/TIMS Joint National Meeting, Nov. 1978, Los Angeles, CA.

Hersey, Paul, and K.H. Blanchard, "The Management of Change," *Training and Development Journal,* Vol. 26, No. 1 (Jan. 1972); Vol. 26, No. 2 (Feb. 1972); and Vol. 26, No. 3 (March 1972).

Hill, R., and B.J. White, *Matrix Organization and Project Management.* Michigan Business Paper #64, University of Michigan, 1979.

Hilton, Joseph R., Jr., *Design Engineering Project Management: A Reference,* Lancaster, PA: Technomic Publishing Company, 1985.

Hirten, J.E., "Managing Capital Projects," *Transportation Quarterly,* July 1984, v38 p.403 (16).

Hlavacik, James D., and Victor A. Thompson, "Bureaucracy and New Product Innovation," *Academy of Management Journal,* 16 (Sep. 1973), pp. 361-372.

Hoadley, P.A., "Mastering the Basics of Project Management (Systems Design)," *Data Management,* March 1987, v25, p.6.

Hoare, H.R., *Project Management Using Network Analysis.* McGraw-Hill, 1973.

Hockney, J.W., *Control and Management of Capital Projects,* Wiley, 1965.

Hodgetts, R.M., "Leadership Techniques in the Project Organization," *Academy of Management Journal 11,* June 1968, pp. 211-219.

Hoge, R.R., "Research and Development Project Management: Techniques for Guiding Technical Programmes Towards Corporate Objectives," *Radio and Electronic Engineer,* 39, Jan. 1970, pp. 211-19.

Holland, Ted, "What Makes a Project Manager?" *Engineering,* 207 (Feb. 14, 1969), p. 262.

Hollander, G.L., "Integrated Project Control, Part II: TCP/Schedule: A Model for Integrated Project Control," *Project Management Quarterly,* June 1973.

Holstein, W.K., "Production Planning and Control Integrated," *Harvard Business Rev,,* May-June, 1968.

Hoos, Ida Russakoff, *Systems Analysis in Public Policy; A Critique,* Berkeley, CA: Univ. of California Press, 1972.

Hopeman, R.J., and D.L. Wilemon, *Project Management/Systems Management-Concepts and Applications*, Syracuse, NY: Syracuse University/NASA, 1973.

Hopeman, Richard J., *Systems Analysis and Operations Management*, Columbus, OH: Merrill, 1969.

Horowitz, J., *Critical Path Scheduling—Management Control Through CPM and PERT*, New York: Roland Press, 1967.

Houre, Henry Ronald, *Project Management Using Network Analysis*, New York: McGraw-Hill, 1973.

Howard, D.C., "Cost/Schedule Control Systems," *Management Accounting*, Oct. 1976.

Howel, R.A., "Multiproject Control," *Harvard Business Rev,*, March-April, 1968.

Hughes, E.R., "Planning: The Essence of Control," *Managerial Planning*, June 1978.

Hynes, Cecil V., "Taking a Look at the Request for Proposal"' *Defense Management Journal*, Oct. 1977, pp. 26-31.

IBM, *Project Management System, Application Description Manual* (H_2O-0210), IBM, 1968.

Ignizio, J.P., *Goal Programming and Extensions*, Lexington Books, 1976.

International Congress on Project Planning by Network *The Practical Application of Project Planning by Network Techniques* (Edited by Mats Ogander), Stockholm: Almgvist & Wiksell; New York: Halsted Press, 1972.

International Congress for Project Planning by Network, *Project Planning by Network Analysis*. Proceedings of the Second International Congress, Amsterdam, the Netherlands, Amsterdam, North-Holland Publishing Company, 1969.

Ivancevich, J., and J. Donnelly, "Leader Influence and Performance," *Personal Psychology*, Vol. 23 (1970), pp. 539-549.

Jackson, B., "Decision Methods for Evaluating R & D Projects," *Research Management*, June-Aug. 1983.

Jacobs, Richard A., "Project Management—A New Style for Success," *S,A,M, Advanced Management Journal*, Vol. 41 (Autumn 1976), pp. 4-14.

Jacobs, Richard A., "Putting Management Into Project Management", Paper presented at A.S.M. Workshops in Detroit, Tulsa, Oakland, and Las Vegas (1976).

Janger, Allen R., "Anatomy of the Project Organization," *Business Management Record*, Nov. 1963, pp. 12-18.

Jansh, Erich, *Design for Evolution; Self-Organization and Planning in the Life of Human Systems*, New York: G. Braziller, 1975.

Jenett, E., "Experience With and Evaluation of Critical Path Methods," *Chemical Engineering*, Feb. 1969.

Jenett, E., "Guidelines for Successful Project Management, *Chemical Engineering* (July 9, 1973), pp. 70-82.

Johnson, Richard Arvid, William T. Newell, and Roger C. Vergin, *Operations Management: A Systems Concept*, Boston: Houghton-Mifflin, 1972.

Johnson, R.A., F.E. Kast, and J.E. Rosenzweig, *The Theory and Management of Systems*, New York: McGraw-Hill, 1973.

Johnson, R.D., "Project Selection and Evaluating," *Long Range Planning*, Sept. 1972.

Johnson, Marvin M., Ed., *Simulation Systems for Manufacturing Industries*, La Jolla, CA: The Society for Simulation, Simulation Councils Inc., t.p. 1973.

Johnson, James R., "Advanced Project Control," *Journal of Systems Management* (May 1977), pp. 24-27.

Jonason, Per, "Project Management, Swedish Style," *Harvard Business Review*, Nov.-Dec. 1971, pp. 104-109.

Jones, Patricia Cukor, Senior Project Manager, Hanscomb Associates, Inc., Atlanta, Ga., personal interview, December 1981.

Kaherlas, H., "A Look at Major Planning Methods: Development, Implementation, Strengths and Limitations," *Long Range Planning*, Aug. 1978.

Kahn, R.L., D.M. Wolfe, R.P. Quinn, J.D. Snock, and R.A. Rosenthal, *Organizational Stress: Studies in Role Conflict ad Ambiguity*, New York: John Wiley, 1964.

Karns, L.A., and L.A. Wanson, "The Effect of Activity Time Variance on Critical Path Scheduling," *Project Management Quarterly*, Dec. 1973.

Kast, Fremont E., *Contingency Views of Organization and Management*, Science Research Associates, 1973.

Kast, Fremont E., *Organization and Management; A Systems Approach*, 2nd Ed. New York: McGraw-Hill, 1974.

Kast, Fremont E., and James E. Rosenzweig, "Organization and management of Space Programs," *On Advances in Space Science and Technology*. Ed. by Frederick I. Ordway, III. New York: Academic Press, 1965.

Kast, D., "The Motivational Basis of Organizational Behavior," *Behavioral Science*, Vol. 9, No. 2, 1964, pp. 131-143.

Keane, A., "Timing for Project Management Control," *Data Management*, 1979.

Kelleher, Grace J., *The Challenge to Systems Analysis: Public Policy and Change*, New York.

Kelley, J., "Critical Path Planning and Scheduling: Mathematical Basis," *Operations Research*, Vol. 9, No. 3, May-June 1961, pp. 296-321.

Kelley, William F., *Management Through Systems and Procedures: A Systems Concept*, New York, 1969.

Kemp, P.S., "Post-Completion Audits of Capital Investment Projects," *Management Accounting*, Aug. 1966.

Kendall, R., "Formulating project methodology," *American City and County*, June 1988, v103 p.14.

Kerridge, Arthur E., and Charles H. Vervalin, Eds., *Engineering and Construction Project Management*, Houston: Gulf Publishing Company, 1986.

Kerzner, H., "Evaluation Techniques in Project Management," *Journal of Systems Management*, Feb. 1980.

Kerzner, Harold, "Systems Management and the Engineer," *Journal of Systems Management*, Oct. 1977, pp. 18-21.

Kerzner, H., "Project Management in the Year 2000," *Journal of Systems Management*, Oct. 1981, v32, p.26 (6).

Kerzner, H., "In Search of Excellence in Project Management," *Journal of Systems Management*, Feb. 1987, v38, p.30 (10).

Kerzner, Harold, *Project Management: A Systems Approach to Planning, Scheduling, and Controlling*, (3rd ed.), New York: Van Nostrand Reinhold, 1989.

Kerzner, Harold, *Project Management for Bankers*, New York: Van Nostrand Reinhold, 1980.

Kerzner, Harold, *Project Management, A Systems Approach to Planning, Scheduling, and Controlling*. New York: Van Nostrand Reinhold Co., 1979.

Khan, M.S., and M.P. Martin, "Managing the Systems Project," *Journal of Systems Management*, Jan. 1989, v40 p.31 (6).

Khatian, G.A., "Computer Project Management—Proposal, Design, and Programming Phases," *Journal of Systems Management*, Aug. 1976.

Kidder, T., *The Soul of a New Machine*. Little, Brown & Co., 1981.

Killian, W.P., "Project Management as an Organizational Concept," *Office*, April 1971, v73, p.14 (4).

Killian, William P., "Project Management—Future Organizational Concepts," *Marquette Business Review*, 2 (1971), pp. 90-107.

Kindred, Alton R., *Data Systems and Management: An Introduction to Systems Analysis and Design*, Englewood Cliffs, N.J.: Prentice Hall, 1973.

King, W.R., and T.A. Wilson, "Subjective Time Estimates in Critical Path Planning—a Preliminary Analysis," *Management Science*, Vol. 13, No. 5, Jan. 1967, pp. 307-320.

Kingdon, Donald R., "The Management of Complexity in a Matrix Organization: A Socio-Technical Approach to Changing Organizational Behavior," Los Angeles: Univ. of California, M.S. Thesis, 1969.

Kingdon, Donald R., *Matrix Organization: Managing Information Technologies*, London: Tavistock Publications, 1973.

Kirchner, Englebert, "The Project Manager," *Space Aeronautics*, 43 (Feb. 1965), pp. 56-64.

Kirkpatick, C.A., and R.C. Levine, *Planning and Control With PERT/CPM, McGraw-Hill, 1966.*

Kliem, Ralph L., The Alexander Hamilton Institute, *The Secrets of Successful Project Management*, New York: Wiley, 1986.

Klimstra, P.D., and J. Potts, "Managing R&D projects," *Research Technology Management*, May/June 1988, v31, p.23 (17).

Klir, George J., *Trends in General Systems Theory*, New York: John Wiley, 1972.

Knill, B., "Project Management: Fastest Lane in the Quest," *Material Handling Engineering*, Sept. 1981, v36, p.86 (5); Oct. 1981, v36, p.84 (5).

Knutson, Joan, and Ira Bitz, *Project Management: How to Plan and Manage Successful Projects*, New York: American Management Association, 1991.

Knutson, J., and M. Scott, "Developing a Project Plan," *Journal of Systems Management*, Oct. 1978.

Koehler, C.T., "Project Planning and Management Technique," *Public Administration Review*, Sept./Oct. 1983, v43, p.459 (8).

Koemtzopoulos, G.A., "Matrix Based Cost Control Systems for the Construction Industry," *Project Management Institute Proceedings*, 1979.

Kondinell, D.A., "Planning Development Projects: Lessons from Developing Countries," *Long Range Planning*, Aug. 1978.

Koplow, Richard A., "From Engineer to Manager—and Back Again," *IEEE Transactions on Engineering Management*, Vol. EM–14, June 1967, pp. 88-92.

Krone, W.T.B., and H.V. Phillips, "SCRAPP, a Reporting and Allocation System for a Multi-Project Situation," *Applications of Critical Path Techniques*, The English Universities Press, Ltd., London, 1968.

Lambertson, L.R., and R.R. Hocking, "Optimum Time Compression in Project Scheduling," *Management Science*, June 1970.

Lammie, J.L., and D.P. Shah, "Project Management: Putting It All Together," *Proceedings of the American Society of Civil Engineers (TE 4 no 15523)*, July 1980, v106, p.437 (15).

Lanford, H.W., and T.M. McCan, "Effective Planning and Control for Large Projects—Using Work Breakdown Structure," *Long Range Planing*, April 1983, v16, p.38 (13).

Langley, R.A. Jr.; and F.E. Meyer, "Managing a Project," *Chemtech*, July 1986, v16, p. 402 (2).

Larsen, S.D., "Control of Construction Projects: An Integrated Approach," *the Internal Auditor*, Sept. 1979.

Larson, Eric. W., D.H. Gobeli, and C.F. Gray, "Application of Project Management by Small Businesses to Develop New Products and Services," *Journal of Small Business Management*, April 1991, v29, p.30 (12).

Lasden, M., "Effective Project Management," *Computer Decisions*, March 1980, v12, p.49 (2).

Laszlo, Ervin, *A Strategy for the Future: The Systems Approach to World Order*. New York: G. Braziller, 1974.

Lawrence, A., "Nuts and Bolts of the Project (Project Management), *Management Today*, Jan. 1991, p.84 (2).

Lazer, R.G., and A.G. Kellner, "Personnel and Organizational Development in an R and D Matrix-Overlay Operation," *IEEE Trans, Eng, Management*, Vol. EM-11, June 1964, pp. 78-82.

Ler, Alec M., *Systems Analysis Frameworks*, New York: John Wiley, 1970.

Levy, J.E., G.L. Thompson, and J.D. Weist, "The ABC's of the Critical Path Method," *Harvard Business Review*, Sept.-Oct. 1963.

Lewis, K., "Group Decision and Social Change," in Maccoby, E.E., et al., *Readings in Social Psychology*, New York; Holt, Rinehart, and Winston, 1958, pp. 197-211.

Lewis, K., "Frontiers in Group Dynamics," *Human Relations*, Vol. 1, No. 1 (1947).

Lewis, James P., *Building and Managing a Winning Project Team*, New York: AMACOM, 1993.

Lewis, James P., *Project Planning, Scheduling and Control*, Chicago: Probus, 1991.

Likierman, A., "Avoiding Cost Escalation on Major Projects," *Management Accounting*, Feb. 1980.

Livingston, G.S., "Weapon System Contracting," *Harvard Business Review*, July-Aug. 1959.

Livingston, J.L., and R. Ronen, "Motivation and Management Control Systems," *Decision Sciences*, April 1975.

Lock, Dennis, *Complete Guide to Project Management*, Boston: Cahners Publishing, 1968.

Lock, Dennis, *Project Management*, Gower Press, 1977.

Lockyer, K.G., *Introduction to Critical Path Analysis*, Pitman Pub. Co., 3rd Ed., 1969, Ch. 1.

Logistics Management Institute, *Introduction to Miliary Program Management*. Washington, D.C.: Sup. of Documents, U.S. Government Printing Office, 1971.

London, Keith R., *The People Side of Systems: The Human Aspects of Computer Systems*, New York: McGraw-Hill, 1976.

Long, J.M., "Applying PERT/CPM to Complex Medical Procedures," *Proceedings of Seminar on Scientific Program Management*, Dept. of Industrial Engineering, Texas A & M University, June 1967.

Lowe, C.W., *Critical Path Analysis by Bar Chart*, London: Business Books, 1966.

Ludwig, Ernest E., *Applied Project Engineering and Management* (2nd Ed.), Houston: Gulf Publishing Company, 1988.

Lustman, F., "Project Management in a Small Organization," *Journal of Systems Management*, Dec. 1983, v34, p.15 (7).

Lutes, Gerald Scott, "Project Selection and Scheduling in the Massachusetts Dept. of Public Works," M.S. Thesis M.I.T. Alfred Sloan School of Management, 1974.

MacCrimmon, K.R., and C.A. Ryavec, "An Analytical Study of the PERT Assumptions," *Operations Research*, Vol. 12, No. 1, Jan.-Feb. 1964, pp. 16-37.

Maciariello, J.A., "Making Program Management Work," *Journal of Systems Management*, July 1974.

Maclead, R.K., "Program Budgeting Works in Non-Profit Institutions," *Harvard Business Review*, Sept. 1971.

Maher, P.M., and A.H. Rubenstein, "Factors Affecting Adoption of a Quantitative Method for R & D Project Selection," *Management Science*, Oct. 1974.

Maieli, Vincent, "Sowing the Seeds of Project Cost Overruns," *Management Review*, 61 (Aug. 1972), pp. 7-14.

Maieli, Vincent, "Management by Hindsight: Diary of a Project Manager," *Management Review*, 60 (June 1971), pp. 4-14.

Maier, N.R., and L.R. Hoffman, "Acceptance and Quality of Solutions as Related to Leader's Attitudes Toward Disagreement in Group Problem Solving," *Journal of Applied Behavioral Science*, 1965, pp. 373-386.

Malcolm, D.G., J.H. Rosenboom, C.E. Clark, and W. Fazar, "Applications of a Technique for R and D Program Evaluation," (PERT), *Operations Research*, Vol. 7, No. 5, 1959, pp. 646-669.

Malloy, J.P., "Computerized Cost System in a Small Plant," *Harvard Business Rev,*, May-June 1968.

Malmquist, D., "Small Project Management: Getting the Project Done on Budget and on Time," *PIMA Magazine*, Nov. 1985, v67, p.20 (5).

Manley, T.R., "Have You Tried Project Management?" *Public Personnel Management*, May 1975, v4, p.180 (9).

Marchbanks, J.L., "Daily Automatic Rescheduling Technique," *Journal of Industrial Engineering*, March 1966.

Marquis, D.C., "A Project Tea, Plus PERT = Success, Or Does It?" *Innovation*, 1969.

Marquis, D.G. and D.M. Straight, Jr., "Organizational Factors in Project Performance," Working Paper No: pp. 133-65, Cambridge M.I.T. School of Management, 1965.

Marshall, A.W., and W.H. Meckling, "Predictability of the Costs, Time, and Success of Development," RAND Corp., Report P-1821, Dec., 1959.

Martin, J.W., "Managing Small R&D Projects—a Learning Model Approach," *Research Management*, May 1980, v23, p.15 (7).

Martin, J.T., *Resource Management*, Wayne, PA: MDI Publications, Management Development Institute, 1968.

Martin, Charles C., *Project Management: How to Make It Work*, New York: Amacom, 1976.

Martin, Charles C., *Project Management: How to Make It Work*, "Applications of a Technique for R and D Program Evaluation," (PERT) AMACOM, New York, 1976, *Operations Research*, Vol. 7, No. 5, 1959, pp. 646-669.

Martin, James Thomas, *Systems Analysis for Data Transmission*, Englewood Cliffs, NJ: Prentice Hall, 1972.

Martin, J., "Planning: The Gap Between Theory and Practice," *Long Range Planning*, Dec. 1979

Martino, R. L., *Project Management*, Wayne, Pa: Management Development Institute, 1968.

Martino, Rocco L. *Project Management and Control*, New York: American Management Association, 1964-1965.

Martyn, A.S., "Some Problems in Managing Complex Development Projects," *Long Range Planning*, April 1975.

Matthies, Leslie H., *The Management System: Systems are People*, New York: John Wiley, 1976.

McCusker, T., "Project Planning Made Easy (Information System Project Management Made Techniques)," *Datamation*, Oct. 15, 1989, v35, p.49 (2).

McGregor, D., *The Professional Manager*, New York: McGraw-Hill, 1967.

McKean, R.N., "Remaining Difficulties in Program Budgeting," In Enke, S., ed., *Defense Management*, Prentice Hall, 1967.

Mclean, Diana, *Project Management Techniques for Performance Monitoring*, The Hague, Netherlands: International Service for National Agricultural Research, 1988.

McMillan, Claud, and Richard F. Gonzalez, *Systems Analysis; A Computer Approach to Decision Models*, Homewood, IL: Irvin, 1973.

McNeely, W.H., "The Importance of Coordination to Project Management," *Supervisory Management*, Nov. 1982, v27, p36 (4).

Mechanic, D., "Sources of Power of Lower Participants in Complex Organizations," *Administrative Science Quarterly*, Vol. 7 (Dec. 1962), pp. 349-364.

Mee, John F., "Project Management," *Business Horizons* 6 Fall 1963, pp. 53-55.

Mee, John F., "Matrix Organization," *Business Horizons*, Summer 1964, p. 70.

Meinhart, W.A., and Leon M. Delionback, "Project Management: An Incentive Contracting Decision Model," *Academy of Management Journal*, Vol. 11 (Dec. 1968), pp. 427-34.

Melchner, Arlyn J., and Thomas A. Kayser, "Leadership Without Formal Authority: The Project Department," *California Management Review*, Vol. 13, No. 2 (1970), pp. 57-64.

Melchner, Arlyn J., Ed., *General Systems and Organization Theory: Methodological Aspects*. Kent, Ohio: Kent University Press, 1975.

Meredith, J., "Program Evaluation Techniques in the Health Services," *American Journal of Public Health*, Nov. 1976.

Merrifield, D.B., "How to Select Successful R & R Projects," *Management Review*, Dec. 1978.

Metzger, Philip W., *Managing a Programming Project* (2nd ed.), Englewood Cliffs, New Jersey: Prentice Hall, 1981.

Meyer, W.C., J.B. Ritter, and L.R. Shaffer, *The Critical Path Method*, McGraw-Hill, 1965.

Middleton, C.J., "How to Set Up a Project Organization," *Harvard Business Review*, 45 (Mar.-April 1967), pp. 73-82.

Miller, E.J., *Systems of Organization*, New York: Barnes and Noble Book Company, 1967.

Miller, R.W., "How to Plan and Control with PERT," *Harvard Bus, Review*, March-April 1962.

Moder, J.J., R.A. Clark, and R.S. Gomez, "Applications of a GERT Simulator to a Repetitive Hardware Development Project," *AIIE Transactions*, Vol. 3, No. 4, 1971, pp. 271-280.

Moder, Joseph John, Cecil R. Phillips, and E.W. Davis, *Project Management with CPM, PERT, and Precedence Diagramming* (3rd Ed.), New York: Van Nostrand Reinhold Company, 1970.

Moder, Joseph John, *Project Management with CPM and Pert*, New York: Van Nostrand Reinhold Publishing, 1964.

Moeller, G.L., and L.A. Digman, "Operations Planning with VERT," *Operation Research*, Vol. 29, No. 4, July-August 1981, pp. 676-697.

Montalbano, M., "High-Speed Calculation of Critical Paths of Large Applications of a Technique for R and D Program Evaluation, (PERT) Networks," IBM Systems Research and Development Center *Technical Operations Research*, Vol. 7, No. 5, 1959, pp. 646-669. Report, Palo Alto, CA, undated report, about 1963.

Moodie, C.L., and D.E. Mandeville, "Project Resource Balancing by Assembly Lines Balancing Techniques," *Journal of Industrial Engineering*, July 1965.

Moore, L.J., and E.R. Clayton, *Introduction to Systems Analysis with GERT Modeling and Simulation*, New York: Petrocelli Books, 1976.

Moore J.R., Jr., and N.R. Baker, "Computational Analysis of Scoring Models for R & D Project Selection," *Management Science*, Dec. 1969.

Moravec, M., "How Organizational Development Can Help and Hinder Project Managers," *Project Management Quarterly*, Sept. 1979.

Mordlea, Irwin, "A Comparison of a Research and Development Laboratory's Organization Structures," *IEEE Transactions on Engineering Management*, Em-14 (Dec. 1967), pp. 170-76.

Morgan, John, "Coping with Resistance to Change," *Ideas for Management*, Cleveland: Assoc. for Systems Management, 1971.

Morreale, R., "Project Planning and Control," *Data Process*, April 1985, v27, p.19 (3).

Morris, L.N., *Critical Path, Construction and Analysis*. Pergammon Press, 1967.

Morrison, R., "Project Management: Organization, Communication, and Keeping the Client Plugged In," *Architectural Record*, November 1982, v170, p.48 (2).

Morton, D.H., "The Project Manager, Catalyst to Constant Change: A Behavioral Analysis," *Project Management Quarterly*, Vol. 6, No. 1 (1975), pp. 22-23.

Moshman, J., J. Johnson, and M. Larsen, "RAMPS—A Technique for Resource Allocation and Multiproject Scheduling," *Proceedings*, Spring Joint Computer Conference, 1963.

Mraz, S., "Simple Project Management for Engineers," *Machine Design*, January 12, 1989, v61 p.147 (2).

Mungo, B.B., "Management Studies in the Field of Aeronautics: Management of Projects," *Journal of the Royal Aeronautical Society*, Vol. 71 (May 1967), pp. 334-336, 336-338 (discussion).

Muth, J.F., and L.A. Swanson, *Industrial Scheduling*, Prentice Hall, 1963.

Myers, S.M., "Conditions for Manager Motivation," *Harvard Business Review*, Jan.-Feb. 1966, pp. 58-71.

Myers, G., "Forms Management; Part 5-How to Achieve Control," *Journal of Systems Management*, Feb. 1977.

Nash, C., and D. Pearce, "Criteria for Evaluating Project Evaluation Techniques," *Journal of the American Institute of Planners*, March 1975.

NATO Institute on Decomposition as a Tool for Solving Large-Scale Problems, Cambridge, England. *Decomposition of Large-Scale Problems.* Amsterdam: North-Holland Publishing Company, 1973.

Neuschel, Richard F., *Management Systems for Profit and Growth*, New York: McGraw-Hill, 1976.

Newman, W.H., *Constructive Control*, Prentice Hall, 1975.

Newnem, A., "Planning Ahead with An Integrated Management Control System," Project Management Proceedings, *1979*,

Newton, J.K., "Computer Modeling for Project Evaluation," *Omega*, May 1981.

Nicholas, J.M., "Successful Project Management: A Force-Field Analysis," *Journal of Systems Management*, Jan. 1989, v40, p.10 (5).

Norko, W.A., "Steps to Successful Project Management (Hardware and Software Conversion Project)," *Journal of Systems Management*, Sept. 1986, v37, p.36 (3).

Nutt, P.C., "Hybrid Planning Methods," *Academy of Management Review*, July 1982.

O'Brien, James Jerome, *CPM in Construction Management: Project Management with CPM.* McGraw-Hill, 1971.

O'Brien, James B., "The Project Manager; Not Just a Firefighter," *S,A,M, Advanced Management Journal*, 39 (Jan. 1974), pp. 52-56.

The Office of the Secretary of Defense and the National Aeronautics and Space Administration, DOD and NASA Guide, *PERT Cost Systems Design*, U.S. Government Printing Office, Washington, D.C., June 1962, Catalog Number D1. 6/2:P94.

Optner, S.L., "Organizational Preference in Research and Development," *Human Relations*, 23 (Aug. 1970), pp. 279-298.

Optner, S. L., *Systems Analysis for Business Management*, Englewood Cliffs, NJ: Prentice Hall, 1975.

Optner, Stanford L., *Systems Analysis for Business Management*, Englewood Cliffs, NJ: Prentice Hall, 1968.

Optner, Stanford L., *Systems Analysis for Business and Industrial Problem Solving*, Englewood Cliffs, NJ: Prentice Hall, 1965.

Oyer, David William, "The Use of Automated Project Management Systems to Improve Information Systems Development," Cambridge, Mass: M.S. Thesis, Alfred P. Sloan School of Management, M.I.T., 1975.

Paige, H.W., "How PERT-Cost Helps the General Manager," *Harvard Business Rev,,* Nov.-Dec. 1963.

Palmer, D., "Project management," *Civil Engineering (London, England)*, Nov./Dec. 1986, p.34.

Paolini, A., Jr., and M.A. Glaser, "Project Selection Methods That Pick Winners," *Research Management*, May 1977.

Parncutt, G., "Concepts of Resource Allocation and Cost Control and Their Utility in Project Management," *Project Management Quarterly*, 1974.

Parris, T.P.E., "Practical Manpower Allocation of a Project Mix Via Zero-Float CPM Networks," *Project Management Institute Proceedings*, 1972.

Pastore, Joseph M., "Organizational Metamorphosis: A Dynamic Model," *Marquette Business Review*, 15 Spring 1971, pp. 17-31.

Patchen, M., *Some Questionnaire Measures of Employee Motivation and Morale: A Report on their Reliability and Validity*, Ann Arbor, MI: Institute for Social Research, 1965.

Patterson, James, "Project Scheduling: The Effects of Problem Structure on Heuristic Performance," *Naval Research Logistics Quarterly*, 1976, Vol. 23, pp. 95.

Patterson, J.H., "Alternate Methods of Project Scheduling with Limited Resources," *Naval Research Logistics Quarterly*, Dec. 1973.

Patterson, James H., and Walter D. Huber, "A Horizon-Varying, Zero-One Approach to Project Scheduling," *Management Science*, Feb. 1974.

Paul, W.J., K. Robertson, and F. Herzberg, "Job Enrichment Pays Off," *Harvard Business Review*, Vol. 7, No. 2 (1969), pp. 61-78.

Pazer, H.H., and L.A. Swanson, *PERTsim, Text and Simulation*, International Textbook Co., 1969.

Pearson, D.A.W., and R. Pearson, "Project Management in Engineering," *Proceedings of the Institution of Mechanical Engineers, Part B, Management and Engineering Manufacture*, 1984, v198, No 6, p.131 (4).

Peart, A.T., *Design of Project Management Systems and Records*, Gower Publishing, 1971.

Peck, M.J., and F.M. Scherer, "The Weapons Acquisition Process: An Economic Analysis," Division of Research, Graduate School of Business Administration, Harvard University, Cambridge, Mass., 1962.

Pegels, C. Carl, *Systems Analysis for Production Operations*, New York: Gordon and Science Publishers, 1976.

Petersen, Perry, "Project Control Systems," *Datamation*, June 1979.

Phillips, Cecil R., "Fifteen Key Features of Computer Programs for CPM and PERT," *Journal of Industrial Engineering*, Jan.-Feb. 1964.

Pinto, J.K., and D.P. Slevin, "Critical Factors in Successful Project Implementation," *IEEE Transactions on Engineering Management*, Feb. 1987, v38, p.7 (23).

Pinto, J.K., and J.E. Prescott, "Planning and Technical Factors in the Project Implementation Process," *Journal of Management Studies*, May 1990, v27, p.305 (23).

Pippin, P.W.T., "Project Management: The Third Discipline in Architectural Practices," *Architectural Record*, June 1981, v169, p.63 (+).

Pitman, B., "Delivering a Product Your Client Really Wants," *Journal of Systems Management*, Feb. 1992, v43, p.18 (1).

Poirot, J.W., "Project Management: Spelling out the Responsibilities," *Consulting-Specifying Engineer*, Nov. 1987, v2, p.92 (4).

Pondy, L.R., "Organizational Conflict: Concepts and Models," *Administrative Science Quarterly*, Sept. 1967, pp. 298-307.

Potter, William J., "Management in the Ad-hocracy," *S,A,M, Advancement Management Journal*, 39 July 1974, pp. 19-23.

Potts, P., "Project Management: Getting Started," *Journal of Systems Management*, Feb. 1982, v33, p.18 (2).

Prince, T., *Information Systems for Management Planning and Control*, Irwin, 1970.

Pritsker, A.A.B., "GERT: Graphical Evaluation and Review Technique," The Rand Corp. RM-4973-NASA, Santa Monica, CA, April, 1966.

Pritsker, A.A.B., and W.W. Harp, "GERT: Graphical Evaluation and Review Technique, Part I. Fundamentals," *Journal of Industrial Engineering*, Vol. 17, No. 5, 1966, pp. 267-274.

Pritsker, A.A.B., and G.E. Whitehouse, "GERT: Graphical Evaluation and Review Technique, Part II. Applications," *Journal of Industrial Engineering*, Vol. 17, No. 5, 1966, pp. 293-301.

Pritsker, A.A.B., "GERT Networks," *The Production Engineer*, Oct. 1968.

Pritsker, A.A.B., *Modeling and Analysis Using Q-GERT Network*, New York: Halsted Press Books (John Wiley & Sons), 1977.

Pritsker, A.A.B., and C. Elliott Sigal, *Management Decision Making: A Network Simulation Approach*, Englewood Cliffs, NJ: Prentice Hall, Inc., 1983.

Pritsker, A.A.B., *The Precedence GERT User's Manual*, Lafayette, IN: Pritsker & Associates, Inc., 1974.

"Project Management Tasks: Wrap Up," *Design News*, April 1982.

Prostick, J.M., "Network Integration, a Tool for Better Management Planning," presented at Meeting of Operations Research Society of America, Philadelphia, Nov. 7-9, 1962.

Putnam, L.H., and A. Fitzsimmons, "Estimating Software Costs," *Datamation*, Oct. 1979, pp. 171-177.

Raithe, A.W., ed., *Gantt on Management*, American Management Association, 1961.

Ramsey, J.E., "Selecting R & D Projects for Development," *Long Range Planning*, Feb. 1981.

Randolph, W. Alan, and Barry Z. Posner, *Effective Project Planning and Management: Getting the Job Done*, Englewood Cliffs, NJ: Prentice Hall, 1988.

Randolph, W.A., and B.Z. Posner, "What Every Manager Needs to Know About Project Management," *Sloan Management Review*, Summer 1988, v29, p.65 (9).

Raudsepp, E., "Build an Effective and Cooperative Team to Ensure Your Project's Success," *EDN*, May 17, 1984, v29, p.305 (2).

Redway, A.K., "Management Tools: Project Planning Procedures," *Industrial Management and Data Systems*, Sept./Oct. 1985, p.7 (5).

Resser, Clayton, "Some Potential Human Problems of the Project Form of Organization," *Academy of Management Journal*, 12 (Dec. 1969), pp. 459-468.

Ricciuti, M., "Easy Way to Manage Multiple IS Projects," *Datamation*, Nov. 15, 1991, v37, p.61 (2).

Rigg, Michael, "Breakthrough Thinking—Improving Project Effectiveness," *Industrial Engineering*, June 1991, v23, p.19 (4).

Ringstrom, N.H., "Making Project Management Work," *Business Horizons*, Fall 1965.

Ritz, George J., *Total Engineering Project Management*, New York: McGraw-Hill, 1990.

Robinson, D.R., "A Dynamic Programming Solution to Cost-Time Tradeoff for CPM," *Management Science*, Oct. 1975.

Rogers, R.A., "Guidelines for Project Management Teams," *Industrial Engineering*, Dec. 1974.

Rolefson, J.F., "Project Management-Six Critical Steps," *Journal of Systems Management*, April 1975.

Roman, D., *R & D Management*, Appleton-Century-Crofts, 1968.

Rosenau, M.D., Jr., "Assessing Project Value," *Industrial Research Development*, May 1979.

Rosenau, M.D., *Successful Project Management*, Wadsworth, 1981.

Rosenau, M.D., Jr., *Successful Project Management*, Lifetime Learning Publications, 1981.

Rubin, Irwin M., and Wychlam Seilig, "Experiences as a Factor in the Selection and Performance of Project Managers," *IEEE Transactions on Engineering Management*, Sept. 1967, pp. 131-135.

Rubinstein, A.H., A.K. Chabrebarty, R.D. O'Keafe, W.E. Souder, and M.C. Young, "Factors Influencing Innovation Success at the Project Level," *Resource, Management,*, 19, 15-20 (May 1976).

Rudwick, Bernard H., *Systems Analysis for Effective Planning: Principles and Cases*, New York: John Wiley, 1969.

Ruskin, Arnold M., and W. Eugene Estes, *What Every Engineer Should Know About Project Management*, New York: M. Dekker, 1982.

Sadler, Philip, "Designing an Organization Structure," *Management International Review*, Vol. 11, No. 6, 1971, pp. 19-33.

Saitow, A.R., "CSPC: Reporting Project Progress to the Top," *Harvard Business Review*, Jan-Feb. 1969.

Sakarev, I., and M. Demirov, *Solving Multi-Project Planning by Network Analysis*, Amsterdam: North-Holland Pub. Co., 1969.

Samaras, Thomas T., *Computerized Project Management Techniques for Manufacturing and Construction Industries,. Englewood Cliffs, NJ: Prentice Hall, 1979.*

Samaras, T.T., "Baseline Management: Control Techniques for New Products," *Supervisory Management*, Dec. 1971, v16 p.7 (6).

Sanders, J., "Effective Estimating Process Outlined," *Computer World*, April 7, 14, and 21, 1980.

Sanderson, M., "Managing the Right Project," *Advanced Management Journal*, Winter 1982, v47, p.59 (4).

Sapolsky, Harvey M., *The Polaris System Development: Bureaucratic and Programmatic Success in Government*, Cambridge, MA: Harvard University Press, 1972.

Sayels, Leonard R., and Margaret K. Chandler, *Managing Large Systems: Organizations for the Future.* New York: Harper and Row, 1971.

Schaller, L.E., *The Change Agent*, New York: Abington Press, 1972.

Scherer, F.M., "Time-Cost Tradeoffs in Uncertain Empirical Research Projects," *Naval Research Logistics Quarterly,* v13, 1, 1966, pp. 71-82.

Schmidt, Joseph William, *Mathematical Foundations for Management Science and Systems Analysis,* New York: Academic Press, 1974.

Schnell, J.S., and R.S. Nicolsi, "Capital Expenditure Feedback: Project Reappraisal," *The Engineering Economist,* Summer 1974.

Schoderbek, Peter P., A.G. Kefalas, and Chagels G. Schoderbek, *Management Systems: Conceptual Considerations,* Dallas: Business Publications, 1975.

Schoderbek, P.P., and L.A. Digman, "Third Generation, PERT/LOB," *Harvard Business Review,,* Sept.-Oct. 1967.

Schonberger, R.J., "Shy Projects Are Always Late: A Rationale Based on Manual Simulation of a PERT/CPM Network," *Interfaces,* Oct. 1981.

Schoof, G., "What Is the Scope of Project Control?" *Project Management Proceedings,* 1979.

Schrage, L., "Solving Resource-Constrained Network Problems by Implicit Enumeration-Non-Preemptive Case," *Operations Research,* 10 (1970).

Schroder, Harold J., "Project Management: Controlling Uncertainty," *Journal of Systems Management,* Vol. 24 Feb. 1975, pp. 28-29.

Schroder, Harold J., "Making Project Management Work," *Management Review,* Vol. 54 (Dec. 1970), pp. 24-28.

Schwartz, S.L., and I. Vertinsky, "Multi-Attribute Investment Decisions: A Study of R & D Project Selection," *Management Science,* Nov. 1977.

Seaton, S.J., "Field Product Performance Reports," *Journal of Systems Management,* Oct. 1978.

Seiler, J.A., "Diagnosing Interdepartmental Conflict," *Harvard Business Review,* Sept.-Oct. 1963.

Selin, G.T.,"Deliver Results with Project Management," *Transportation and Distribution,* May 1991, v32, p38 (3).

Sethi, N.K., "Project Management," *Industrial Management*, Jan.-Feb. 1980.

Shah, Ramesh P., "Project Management: Cross Your Bridges Before You Come to Them," *Management Review*, 60 (Dec. 1971), pp. 21-27.

Shaheen, Salem K., *Practical Project Management*, New York: Wiley, 1987.

Shannon, Robert E., "Matrix Management Structures," *Industrial Engineering*, 4 (March 1972), pp. 26-29.

Sharad, D., "About Delays, Overruns and Corrective Actions," *Project Management Quarterly*, Dec. 1976, pp. 21-25.

Sheriff, M., "Superordinate Goals in the Reduction of Intergroup Conflict," *American Journal of Sociology*, No. 63 (1958), pp. 349-358.

Shih, W., "A Branch and Bound Procedure for a Class of Discrete Resource Allocation Problems with Several Constraints," *Operational Research Quarterly*, June 1977.

Shih, W., "A New Application of Incremental Analysis in Resource Allocations," *Operational Research Quarterly*, Dec. 1974.

Shranks, J.G., "Managing Projects Requires four Main Ingredients," *Data Management*, Dec. 1983, v21 p.14 (2).

Shrode, William A., and Dan Voich, Jr., *Organization and Management: Basic Systems Concepts*, Homewood, IL: R.D. Irwin, 1974.

Shull, Fremont A., *Matrix Structure and Project Authority for Optimizing Organizational Capacity*, Business Science Monograph No. 1, Carbondale, Business Research Bureau, Southern Illinois University, 1965.

Shull, Fremont, and R.J. Judd, "Matrix Organizations and Control Systems," *Management International Review* 11, No. 6 (1971), pp. 65-72.

Silverman, Melvin, *Project Management: A Short Course for Professionals*, Wiley Publishers, 1976.

Silverman, M., *Project Management: A Short Course for Professionals*, Wiley, 1976.

Silverman, Melvin, *Project Management: A Short Course for Professionals* (2nd ed.), New York: Wiley, 1988.

Simmons, John R., *Management of Change: The Role of Information*, based on a research project sponsored by the Institute of Office Management, London: Gee & Co., 1970.

Simpson, Dwain W., *New Techniques in Software Project Management*, New York: Wiley, 1987.

Sivazlian, B.D., and L.E. Stanfeld, *Analysis of Systems in Operations Research*. Englewood Cliffs, NJ: Prentice Hall, 1973.

Smith, Michael, *PCS: A Project Control System*, Thesis, M.I.T., Cambridge, MA, 1973.

Smith, G.A., "Program Management—Art or Science?" *Mechanical Engineering*, Vol. 96 (Sept. 1974), pp. 18-22.

Smith, Larry A., and Peter Mahler, "Comparing Commercially Available CPM/PERT Computer Programs," *Industrial Engineering*, April 1978.

Smith, Peter, *Agricultural Project Management: Monitoring and Control of Implementation*. London; New York: Elsevier Applied Science Publishers, 1984.

Smith, W.P., "Successful Energy Project Management," *Plant Engineering*, August 9, 1984, v38 p.64 (4).

Smyster, Craig H., "A Comparison of the Needs of Program and Functional Management," unpublished Masters Thesis, School of Engineering, Wright-Patterson Air Force Base, Air Force Institute of Technology, 1965.

Snodgrass, Raymond J., *The Concept of Project Management*, Washington, DC: U.S. Army Material Command Historical Office, 1964.

Snowdon, M., "Measuring Performance in Capital Project Management," *Long Range Planning*, Aug. 1980.

Souder, W.E., "Utility and Perceived Acceptability of R & D Project Selection Models," *Management Science*, Aug. 1973.

Souder, W.E., "System for Using R & D Project Evaluation Methods," *Research Management*, Sept. 1978.

Souder, W.E., "Autonomy, Gratification, and R & D Outputs: A Small Sample Field Study," *Management Science*, April 1974.

Souder, W.E., "Analytical Effectiveness of Mathematical Models for R & D Project Selection," *Management Science*, April 1973.

Spinner, M., *Elements of Project Management: Plan, Schedule, and Control*, Prentice Hall, 1981.

Stallworthy, E.A. and Om P. Kharbanda, *Project and Company Management: The Road to the Top*, Bradford, England: MCB University Press LTD, 1988.

Steiner, George Albert, *Industrial Project Management*, Macmillan, 1968.

Staples, E., "Managing a Systems Project," *Office Administration Automation*, June 1984, v45, p. 80.

Staropli, G.K., "Project Management: Controlling Uncertainty," *Journal of Systems Management*, May 1975, v26 p.28 (2).

Starr, Martin Kenneth, *Production Management: Systems and Synthesis*, 2nd ed., Englewood Cliffs, NJ: Prentice Hall, 1972.

Steger, W.A., "How to Plan for Management in New Systems," *Harvard Business Review*, Sept.-Oct. 1962.

Steiner, V.M., "Computer Aided Project Management," *Plant Engineering*, January 21, 1988, v4,2 p.50 (11).

Stewart, J.M., "Guides to Effective Project Management," *Management Review*, Jan. 1966.

Stewart, John M., "Making Project Management Work," *Business Horizons*, 8 (Fall 1965), pp. 54-68.

Stillman, William James, *Construction Practices for Project Managers and Superintendents*. Reston, Va.: Reston Publishing Company, 1978.

Stinson, Joel P., E.W. Davis, and B. Khumawala, "Multiple Resource-Constrained Scheduling Using Branch and Bound," *AIIE Transactions*, Sept. 1978.

Stopher, Peter R., and Arnim H. Meyburg, *Transportation Systems Evaluation*, Lexington, MA: Lexington Books, 1976.

Strafford, J.E., "Construction Project Management," *Accountancy*, Sept. 1977, v88 p.56 (+).

Strasch, Stanley F., *Systems Analysis for Marketing Planning and Control*, Glenview, IL: Scott, Foresman, 1972.

Strauss, T., "The Articulation of Project Work: An Organizational Process," *Sociological Quarterly*, Summer 1988, v29 p.163 (16).

Stuckenbruck, L.C., *The Implementation of Project Management*, Reading, MA: Addison-Wesley, 1981.

Stuckenbruck, Linn C., Ed., *The Implementation of Project Management: The Professional's Handbook*, Reading, MA: Addison-Wesley Publishing Company, 1981.

Sullivan, C.A., "How to Get the Job Done," *Journal of Systems Management*, Feb. 1992, v43, p.17 (1).

Sunage, T., "A Method of the Optimal Scheduling for a Project with Resource Restrictions," *Journal of the Operations Research Society of Japan*, March 1970.

Talbot, F.B., "Project Scheduling with Resource-Duration Interactions: The Nonpreemptive Case," Working paper No. 200, Graduate School of Business Administration, University of Michigan, Jan. 1980.

Talbot, F.B., and J.H. Patterson, "An Efficient Integer Programming Algorithm With Network Cuts for Solving Resource-Constrained Scheduling Problems," *Management Science*, July 1978.

Talbot, B.F., and J.H. Patterson, "Optimal Methods for Scheduling Under Resource Constraints," *Project Management Quarterly*, Dec. 1979.

Tannenbaum, Robert, and Warren H. Schmidt, "How to Choose a Leadership Pattern," *HBR Classic* (May-June 1973), pp. 162-180.

Taylor, William John, *Practical Project Management*. New York: Wiley 1973.

Taylor, William John, and T.F. Watling, *Successful Project Management*, London: Business Books, 1970.

Thomsett, Michael C., *The Little Black Book of Project Management*, New York: American Management Association, 1990.

Taylor, W.J., "Teamwork Through Conflict," *Business Week* (March 20, 1971), pp. 44-45.

Thamhain, Hans J., and Gary R. Gemmill, "Influence Styles of Project Managers: Some Project Performance Correlates," *Academy of Management Journal*, Vol. 17 (June 1974), pp. 216-224.

Thamhain, Hans J., and David L. Wilemon, "Diagnosing Conflict Determinants in Project Management," *IEEE Transactions on Engineering Management*, Vol. EM-22 (Feb. 1975), pp. 35-44.

Thamhain, Hans J., and David L. Wilemon, "Conflict Management in Project-Oriented Work Environments," *Proceedings of the Sixth International Meeting of the Project Management Institute*, Washington, D.C., Sept. 18-21, 1974.

Thamhain, Hans J., "The Effective Management of Conflict in Project-Oriented Work Environments," *Defense Management Journal*, Vol. 11, No. 3, July 1972, p. 975.

Thamhain, Hans J., "Conflict Management in Project Life Cycles," *Sloan Management Review* (Summer 1975), pp. 31-50.

Thompson, J.D., *Organization in Action*, New York: McGraw-Hill, 1967.

Thompson, Victor A., "Bureaucracy and Innovation," *Administrative Science Quarterly*, Vol. 10, June 1965, pp. 1-20.

Thornberry, N.E., "Training the Engineer as a Project Manager," *Training and Development Journal*, Oct. 1987, v41 p.60 (3).

Toellner, J.D., "Project Management: A Formula for Success," *Computer World*, Dec. 1978.

Toellner, John, "Project Estimating," *Journal of Systems Management*, May 1977, pp. 6-9.

Tonge, F.M., *A Heuristic Program for Assembly Line Balancing*, Prentice Hall, 1961.

Trower, Michael H., "Fast Track to Project Delivery: Systems Approach to Project Management," *Management Review*, Vol. 62, April 1973, pp. 19-23.

Tsai, Martin Chia-Ping, "Contingent Conditions for the Creation of Temporary Management Organizations," M.S. Thesis, Alfred P. Sloan School of Management, M.I.T. Cambridge, MS, 1976.

Turban, Efraim, "The Line of Balance—A Management by Exception Tool," *The Journal of Industrial Engineering*, Vol. 19, No. 9, Sept. 1968, pp. 440-448.

Turner, W.S., III, *Project Auditing Methodology*, Amsterdam: North Holland Publishing Co., 1980.

Ulgen, U.M., "Project Management Techniques Are the Key to Successful Simulation Projects," *Industrial Engineering*, Aug. 1991, v23, p.37 (5).

Vanston, J.H., Jr., "Use of the Partitive Analytical Forecasting (PAF) Technique for Analyzing of the Effects of Various Funding and Administrative Strategies on Nuclear Fusion Power Plan Development," University of Texas, TR ESL-15, Energy Systems Laboratory, 1974.

Vaughan, Dennis Henry, "Key Variables of a Management Information System for a Department of Defense Project Manager," M.S. Thesis, Alfred P. Sloan School of Management, M.I.T., Cambridge, MS, 1976.

Vaughan, D.H., "Understanding Project Management, *Manage*, Vol. 19, No. 9, 1967, pp. 52-58.

Vaugn, C.F., "Project Management: No More Surprises," *Correct Today*, April 1986, v48 p.80 (2).

Vernon, P., "Managing a Major Project," *Civil Engineering (American Society of Civil Engineers)*, July/August 1988, p.53 (+),

Wadsworth, M., *EDP Project Management Controls*, Englewood Cliffs, NJ: Prentice Hall, 1972.

Walton, R.E., J.M. Dutton, and T.P. Cafferty, "The Management of Interdepartmental Conflict: A Model and Review," *Administrative Science Quarterly*, Vol. 14, No. 1, March 1969, pp. 78-84.

Wadsworth, M.D., *EDP Project Management Controls*, Prentice Hall 1972.

Waldrop, J.H., "Project Management: Have We Applied All We Know?" *Information and Management*, Feb. 1984, v7, p.13 (8).

Walker, M.R., and J.S. Sayer, "Project Planning and Scheduling," Report 6959, E.I. du Pont de Nemours and Co., Wilmington, Delaware, March 1959.

Walker, Anthony. Ph.D., *Project Management in Construction*, London; New York: Granada, 1984.

Wall, W.C., Jr., "Ten Proverbs of Project Control," *Research Management*, March 1982, v25, p.26 (4).

Wallace, D., "Get It Done! Project Management—Your Most Valuable Tool," *Success*, March 1990, v37, p.46 (2).

Walsh, J.J., and J. Kanter, "Toward More Successful Project Management," *Journal of Systems Management*, Jan. 1988, v39 p.16 (6).

Walton, Dutton, and Cafferty, "Organizational Contest and Interdepartmental Conflict," *Administrative Science Quarterly*, Vol. 14, No. 4, March 1969, pp. 522-542.

Webb, James E., "NASA as an Adaptive Organization," *On Technological Change and Management*, ed. by David W. Ewing, Cambridge, MA: Harvard Univ. Press, 1970.

Weber, F.M., "Ways to Improve Performance on Projects," *Project Management Quarterly*, Sept. 1981.

Webster, F.M., Jr., *Survey of Project Management Software Packages*, Project Management Institute, Drexel Hill, PA, 1982.

Wedley, W.C., and A.E.J. Ferrie, "Perceptual Differences and Effects on Managerial Participation on Project Management," *Operations Research*, March 1978.

Weinberg, Gerald M., *An Introduction to General Systems Thinking*, New York: John Wiley, 1975.

Weist, Jerry D., "Precedence Diagramming Methods: Some Unusual Characteristics and Their Implications for Project Managers," Journal of Operations Management, Vol. 1, No. 3, Feb. 1981, pp. 121-130.

Weist, Jerome D., "Heuristic Programs for Decision Making," *Harvard Business Review*, Sept.-Oct. 1965.

Weist, Jerome D., "Heuristic Model for Scheduling Large Projects with Limited Resources," Management Science, February 1967.

Weist, Jerome D., and Ferdinand K. Levy, *A Management Guide to PERT/CPM*. Englewood Cliffs, NJ: Prentice Hall, Inc., 1969.

West, O.E., et al., *Survey of CPM Scheduling Software Packages and Related Project Control Programs*, Project Management Institute, Drexel Hill, Pennsylvania, January 1980.

Westmore, P.J., "Organizing and Controlling the Big Capital Project," *Director*, Jan. 1972, v24, p.77 (4).

Westney, Richard E., *Managing the Engineering and Construction of Small Projects: Practical Techniques for Planning, Estimating, Project Control, and Computer Applications*, New York: Dekker, 1985.

Wetzel, John Jay, "Project Control at the Managerial Level in the Automotive Engineering Environment," M.S. Thesis, Alfred Sloan School of Management, M.I.T., Cambridge, MS, 1973.

Whaley, W.M., and R.A. Williams, "A Profits-Oriented Approach to Project Selection," *Research Management*, Sept. 1969.

Wheelwright, S.C., and R.L. Blank, "Involving Operating Managers in Planning Process Evaluation," *Sloan Management Review*, Summer 1979.

Wheelwright, S.C., and K.B. Clark, "Creating Project Plans to Focus Product Development," *Harvard Business Review*, March-April 1992, v70, p.70 (13).

White, Anthony G., *Project Management, Pulling it all Together: A Selected Bibliography*. Monticello, IL: Vance Bibliographies, 1979.

Whitehouse, Gary E., *Systems Analysis and Design Using Network Techniques*. Englewood Cliffs, NJ: Prentice Hall, 1973.

Whitehouse, Gary E., "Project Management Techniques," *Industrial Engineering*, Vol. 5 (March 1973), pp. 24-29.

Whiting, Richard J., "In Defense of Functional Organization," *Management Review*, Vol. 58, No. 7, July 1969, pp. 49-52.

Whitley, F., "Planning and Implementing Capital Improvement Projects," *Training and Development*, Sept. 1991, v106, p.28 (2).

Wilemon, David L., "Managing Conflict in Temporary Management Systems," *Journal of Management Studies*, Vol. 10, Oct. 1973, pp. 282-296.

Wilemon, D.L., "Managing Conflict on Project Teams," *Management Journal*, Summer 1974, pp. 28-34.

Wilemon, D.L., "Project Management Conflict: A View from Apollo," *Third Annual Symposium of the Project Management Institute*, Houston, Texas, October, 1971.

Wilemon, D.L. and Gary R. Gemmill, "Interpersonal Power in Temporary Management Systems," *Journal of Management Studies*, 8 (Oct. 1971), pp. 315-328.

Wilemon, D.L., and John P. Cicero, "The Project Manager: Anomalies and Ambiguities," *Academy of Management Journal*, Vol. 13 (Sept. 1970), pp. 269-282.

Wilemon, D.L., "Project Management and its Conflicts: A View from Apollo," *Chemical Technology*, Vol. 2, No. 9, Sept. 1972, pp. 527-534.

Williams, D.J., "A Study of a Decision Model for R & D Project Selection," *Operational Research Quarterly*, Sept. 1969.

Willoughby, T.C. and J.A. Senn, *Business Systems*, Cleveland: Association for Systems Management, 1975.

Willoughby, Theodore C., *Business Systems*, Cleveland: Association for Systems Management, 1975.

Wilson, Ira Gaulbert, *Management Innovation and System Design*, Princeton, NJ: Auerbach, 1971.

Wolfe, P.M., "Using GERT as a Basis for a Project Management System," *ORSA/TIMS Joint National Meeting*, Los Angeles; CA, Nov. 1978.

Wolff, M.F., "When Projects Select You or the Researcher as a Fire Fighter," *Research Management*, May/June, 1983 v26 p.8 (2).

Wolff, W.F., "Rules of Thumb for Project Management," *Research Management*, July/August 1984, v27 p.11 (3).

Woodgate, Harry Samuel, *Planning by Network: Project Planning and Control Using Network Techniques* (2nd ed.), London: Business Publications, 1967.

Woodgate, Harry Samuel, *Planning by Network: Project Planning and Control Using Network Techniques*, London: Business Publications, 1967.

Woodworth, Bruce M., and C.T. Willie, "A Heuristic Algorithm for Resource Levelling in Multi-Project, Multi-Resource Scheduling," *Decision Sciences*, 1975, Vol. 6, pp. 525-540.

Wooldridge, Susan, *Project Management in Data Processing* (1st Ed.), New York: Petrocelli/Charter, 1976.

Wright, N.H., "Matrix Management, a Primer for the Administrative Manager," *Management Review*, May 1979.

Wrong, D., "Some Problems in Defining Social Power," *American Journal of Sociology*, Vol. 73, No. 6, (May 1968), pp. 673-681.

Youker, Robert, "Organization Alternatives for Project Managers," *Management Review*, November 1977.

Zaloon, V.A., "Project Selection Methods," *Journal of Systems Management*, Aug. 1973.

Zeldman, M., *Keeping Technical Projects on Target*, American Management Association, 1978.

References

Aaker, David A. *Developing Business Strategies. New York: Wiley, 1984.*

Adams, James L. Conceptual Blockbusting: *A Guide to Better Ideas,* second edition. New York: W.W. Norton, 1979.

Adams, John D. (Editor) *Transforming Leadership: From Vision to Results.* Alexandria, VA: Miles River Press, 1986.

Adams, John D. (Editor) *Transforming Work.* Alexandria, VA: Miles River Press, 1984.

Ailes, Roger. *You Are the Message: Secrets of the Master Communicators.* Homewood, IL: Dow Jones-Irwin, 1988.

Archibald, R.D., and Villoria, R.L. *Network-Based Management Systems (PERT/CPM)*. New York: Wiley, 1967.

Argyris, Chris. *Overcoming Organizational Defenses: Facilitating Organizational Learning*. Boston: Allyn and Bacon, 1990.

Axelrod, Robert. *The Evolution of Cooperation*. New York: Basic Books, 1984.

Beer, Stafford. *Brain of the Firm*, second edition. Chichester, England: Wiley, 1981.

Beer, Stafford. *The Heart of Enterprise*. Chichester, England: Wiley, 1979.

Bennis, Warren G.; Benne, Kenneth D.; Chin, Robert; and Corey, Kenneth E. *The Planning of Change*, third edition. New York: Holt, Rinehart and Winston, 1976.

Bennis, Warren G., and Nanus, Burt. *Leaders: The Strategies for Taking Charge*. New York: Harper & Row, 1985.

Benveniste, Guy. *Mastering the Politics of Planning*. San Francisco: Jossey-Bass, 1989.

Blake, Robert E., and Mouton, Jane S. *The Managerial Grid*. Houston, TX: Gulf Publishing, 1964.

Blanchard, Benjamin S. *Engineering Organization and Management*. Englewood Cliffs, NJ: Prentice-Hall, 1976.

Block, Peter. *The Empowered Manager*. San Francisco: Jossey-Bass, 1987.

de Bono, Edward. *New Think*. New York: Avon Books, 1971.

de Bono, Edward. *Serious Creativity*. New York: Harper, 1992.

de Bono, Edward. *Six Thinking Hats*. Boston: Little, Brown & Co., 1985.

Burrill, Claude W., & Ellsworth, Leon W. *Modern Project Management*. Tenafly, NJ: Burrill-Ellsworth Assoc., 1980.

Burt, D.K.; Norquist, W.E.; & Anklesaria, J. *Zero Base Pricing: Achieving World Class Competitiveness Through Reduced All-in-Costs*. Chicago: Probus, 1990.

Busch, D.H. *The New Critical Path Method: The State-of-the-Art In Project Modeling and Time Reserve Management.* Chicago: Probus, 1989.

Calero, Henry H., & Oskam, Bob. *Negotiate the Deal You Want.* New York: Dodd, Mead & Company, 1983.

Carlzon, Jan. *Moments of Truth.* New York: Perennial, 1987.

Cleland, David I., and King, William R., Editors. *Project Management Handbook.* New York: Van Nostrand Reinhold, 1983.

Cleland, David I., and King, William R. *Systems Analysis and Project Management,* Second edition. New York: McGraw-Hill, 1968, 1975.

Cohen, Herb. *You Can Negotiate Anything.* New York: Bantam, 1980.

Coxon, R., "How Strategy Can Make Major Projects Prosper". *Management Today.* April 1983.

Curtis, Dan B.; Mazza, J.M.; and Runnebohm, S. *Communication for Problem Solving.* New York: Wiley, 1979.

Davis, Philip J., & Park, David. *No Way: The Nature of the Impossible.* New York: W. H. Freeman, 1987.

Davis, Stanley M., and Lawrence, Paul R. *Matrix.* Reading, MA: Addison-Wesley, 1977.

Deal, T.E., and Kennedy, A.A. *Corporate Cultures: The Rites and Rituals of Corporate Life.* Reading, MA: Addison-Wesley, 1982.

Deming, W.E. *Out of the Crisis.* Cambridge, MA: MIT Press, 1982.

Dimancescu, Dan. *The Seamless Enterprise: Making Cross Functional Management Work.* New York: Harper, 1992.

Drucker, Peter F. *Innovation and Entrepreneurship.* New York: Harper & Row, 1985.

Drucker, Peter F. *Management: Tasks, Responsibilities, Practices.* New York: Harper & Row, 1973, 1974.

Drucker, Peter F. *The Practice of Management.* New York: Harper, 1954, 1956.

Dyer, William G. *Team Building: Issues and Alternatives.* Reading, MA: Addison-Wesley, 1977.

Fleming, Q.W. *Cost/Schedule Control Systems Criteria: The Management Guide To C/SCSC.* Chicago: Probus, 1988.

Fleming, Q.W.; Bronn, J. W.; and Humphreys, Gary C. *Project and Production Scheduling.* Chicago: Probus, 1987.

Fleming, Q.W., & Fleming, Q.J. *Subcontract Project Management: Progress Payments.* Chicago: Probus, 1992.

Fournies, Ferdinand F. *Coaching for Improved Work Performance.* New York: Van Nostrand, 1978.

Foster, Richard. *Innovation: The Attacker's Advantage.* New York: Summit Books, 1986.

Francis, Dave, and Young, Don. *Improving Work Groups: A Practical Manual for Team Building.* San Diego, CA: University Associates, 1979.

Hackman, J. Richard, and Oldham, Greg R. *Work Redesign.* Reading, MA: Addison-Wesley, 1980.

Hanna, David P. *Designing Organizations for High Performance.* Reading, MA: Addison-Wesley, 1988.

Harman, Willis, & Rheingold, Howard. *Higher Creativity: Liberating the Unconscious for Breakthrough inSights.* Los Angeles: Jeremy P. Tarcher, Inc., 1984.

Harvard Business Review of Management. New York: Harper & Row, 1975.

Harvard Business Review on Human Relations. New York: Harper & Row, 1979.

Harvey, Jerry B. *The Abilene Paradox: And Other Meditations on Management.* San Diego: University Associates, 1988.

Hersey, Paul. *The Situational Leader.* New York: Warner Books, 1984.

Hersey, Paul, and Blanchard, Kenneth. *Management of Organizational Behavior: Utilizing Human Resources,* Fourth edition. Englewood Cliffs, NJ: Prentice-Hall, 1981.

Juran, J.M. *Leadership for Quality.* New York: Free Press, 1989.

Kanter, Rosabeth M. *The Change Masters.* New York: Simon & Schuster, 1984.

Kerzner, Harold. *Project Management: A Systems Approach to Planning, Scheduling, and Controlling.* New York: Van Nostrand, 1979.

Kouzes, James M., and Posner, Barry Z. *The Leadership Challenge: How to Get Extraordinary Things Done in Organizations.* San Francisco: Jossey-Bass, 1987.

Kuhn, Thomas S. *The Structure of Scientific Revolutions,* 2nd Ed. Chicago: University of Chicago Press, 1970.

Laborde, Genie Z. *Influencing with Integrity.* Palo Alto, CA: Syntony, 1984.

Lassey, W.R., and Sashkin, M. (Editors). *Leadership and Social Change,* third edition. San Diego: University Associates, 1983.

Lawler, E.E., III. *High-Involvement Management.* San Francisco: Jossey-Bass, 1986.

Lax, David A., and Sebenius, James K. *The Manager as Negotiator.* New York: The Free Press, 1986

Levitt, Theodore. *The Marketing Imagination.* New York: The Free Press, 1983.

Lynch, Dudley, and Kordis, Paul. *Strategy of the Dolphin: Scoring a Win in a Chaotic World.* New York: William Morrow & Company, 1988.

Maccoby, Michael. *Why Work: Leading the New Generation.* New York: Simon and Schuster, 1988.

Mali, Paul, Editor. *Management Handbook.* New York: Wiley, 1981.

Maciariello, Joseph A. *Program-Management Control Systems.* New York: Wiley, 1978.

March, James G., and Simon, Herbert A. *Organizations.* New York: Wiley, 1958.

McClelland, David. *Power: The Inner Experience.* New York: Irvington, 1975.

Mendelssohn, Kurt. *The Riddle of the Pyramids*. New York: Praeger, 1974.

Meredith, Jack R., and Mantel, Jr., Samuel J. *Project Management: A Managerial Approach*. New York: Wiley, 1985.

Miller, William C. *The Creative Edge: Fostering Innovation Where You Work*. Reading, MA: Addison-Wesley, 1986.

Mintzberg, Henry. *Mintzberg on Management: Inside our Strange World of Organizations*. New York: The Free Press, 1989.

Moder, Joseph J., Phillips, Cecil R., and Davis, Edward W. *Project Management with CPM, PERT, and Precedence Diagramming*, Third edition. New York: Van Nostrand, 1983.

Nadler, Gerald, & Hibino, Shozo. *Breakthrough Thinking*. Rocklin, CA: Prima Publishing, 1990.

Nierenberg, Gerard I. *The Complete Negotiator*. New York: Nierenberg & Zeif, 1986.

von Oech, Roger. *A Whack on the Side of the Head*. New York: Warner, 1983.

von Oech, Roger. *A Kick in the Seat of the Pants*. New York: Warner, 1986.

Oncken, Jr., William. *Managing Management Time*. Englewood Cliffs, NJ: Prentice-Hall, 1984.

Pasmore, W.A. *Designing Effective Organizations: The Sociotechnical Systems Perspective*. New York: Wiley, 1988.

Pava, Calvin. *Managing New Office Technology*. New York: Free Press, 1983.

Perlman, K.I. *Handbook of Purchasing and Materials Management*. Chicago: Probus, 1990.

Peters, Tom. *Thriving on Chaos*. New York: Alfred A. Knopf, 1987.

Pfeiffer, J. W.; Goodstein, L. D.; & Nolan, T. M. *Understanding Applied Strategic Planning: A Manager's Guide*. University Associates, 8517 Production Ave., San Diego, CA, 92121, 1985.

Project Management Journal. Project Management Institute, P.O. Box 43, Drexel Hill, PA 19026.

Ray, M., & Myers, R. *Creativity in Business.* Garden City, NY: Doubleday, 1986.

Reddy, W.B., and Jamison, K. (Editors). *Team Building: Blueprints for Productivity and Satisfaction.* Alexandria, VA: National Training Labs, 1988.

Rickards, Tudor. *Problem solving through creative analysis.* Epping, Essex, England: Gower Press, 1975.

Rifkin, Jeremy. *Time Wars.* New York: Touchstone, 1987.

Robert, Marc. *Managing Conflict from the Inside Out.* San Diego: University Associates, 1982.

Rosenblatt, Alfred, & Watson, George F. *Concurrent Engineering.* IEEE Spectrum, July 1991.

Savage, C.M. *Fifth Generation Management.* Bedford, MA: Digital Press, 1990.

Scholtes, Peter R. *The Team Handbook.* Madison, WI: Joiner Associates, 1988.

Schumacher, E.F. *Small is Beautiful: Economics as if People Mattered.* New York: Perennial Library, 1989 (reprint).

Senge, Peter. *The Fifth Discipline.* New York: Doubleday, 1990.

Seymour, D.T., Editor. *The Pricing Decision: A Strategic Planner for Marketing Professionals.* Chicago: Probus, 1989.

Slevin, Dennis P. *Executive Survival Manual: A Program for Managerial Effectiveness.* Pittsburgh, PA: Innodyne, P.O. Box 111386 (Zip 15238), 1985.

Stewart, Rodney. *Cost Estimating,* Second Edition. New York: Wiley, 1991.

Toffler, Alvin. *Future Shock.* New York: Bantam Books, 1971.

Vroom, Victor H., & Jago, Arthur G. *The New Leadership: Managing Participation in Organizations.* Englewood Cliffs, NJ: 1988.

Walpole, Ronald E. *Introduction to Statistics*, Second Edition. New York: Macmillan, 1974.

Walton, Richard E. *Interpersonal Peacemaking: Confrontations and Third-party Consultation*. Reading, MA: Addison-Wesley, 1969.

Waterman, Robert H. *The Renewal Factor*. New York: Bantam, 1987.

Watzlawick, P.; Beavin, J.; and Jackson, D. *Pragmatics of Human Communication*. New York: Norton, 1967.

Weisbord, Marvin R. *Productive Workplaces*. San Francisco: Jossey-Bass, 1987.

Winner, Robert I.; James P. Pennell; Harold E. Bertrand; & Marko M.G. Slusarczuk. *The Role of Concurrent Engineering in Weapons System Acquisition*. Alexandria, VA: Institute for Defense Analysis, 1988.

Winston, Stephanie. *The Organized Executive*. New York: Warner Books, 1983.

Index